NANOSCALE TRANSISTORS

Device Physics, Modeling and Simulation

T0138062

NANOSCALE TRANSISTORS

Device Physics, Modeling and Simulation

Mark S. Lundstrom
Purdue University
West Lafayette, Indiana
USA

Jing Guo
University of Florida
Gainesville, Florida
USA

 Springer

Mark S. Lundstrom
Purdue University
West Lafayette, IN, USA

Jing Guo
University of Florida
Gainesville, FL, USA

Nanoscale Transistors: Device Physics, Modeling and Simulation

ISBN 10: 0-387-28003-3 (e-book)
ISBN 13: 978-1-4419-3915-9   ISBN 13: 978-0-387-28003-5 (e-book)

Printed on acid-free paper.

springeronline.com

Table of Contents

Preface

Silicon technology continues to progress, but device scaling is rapidly taking the metal oxide semiconductor field-effect transistor (MOSFET) to its limit. When MOS technology was developed in the 1960's, channel lengths were about 10 micrometers, but researchers are now building transistors with channel lengths of less than 10 nanometers. New kinds of transistors and other devices are also being explored. Nanoscale MOSFET engineering continues, however, to be dominated by concepts and approaches originally developed to treat microscale devices. To push MOSFETs to their limits and to explore devices that may complement or even supplant them, a clear understanding of device physics at the nano/molecular scale will be essential. Our objective is to provide engineers and scientists with that understanding – not only of nano-devices, but also of the considerations that ultimately determine system performance. It is likely that nanoelectronics will involve much more than making smaller and different transistors, but nanoscale transistors provides a specific, clear context in which to address some broad issues and is, therefore, our focus in this monograph.

This monograph was written for engineers and scientists who are engaged in work on nanoscale electronic devices. Familiarity with basic semiconductor physics and electronics is assumed. Chapter 1 reviews some central concepts, and Chapter 2 summarizes the essentials of traditional semiconductor transistors, digital circuits, and systems. This material provides a baseline against which new devices can be assessed. At the same time, it defines the requirements of a device for it to be useful in a digital electronic system. Chapter 3 presents a nontraditional view of the ballistic MOSFET. By treating a traditional device from a fresh perspective, this chapter introduces electrical engineers to new ways of thinking about small electronic devices. In Chapter 4, we extend the model to discuss the physics of scattering in nanotransistors. Chapter 5 uses the same, general approach to treat semiconductor nanowire and carbon nanotube FETs. Finally, in Chapter 6, we introduce a 'bottom-up' view by discussing electronic conduction in molecules and showing how a simple model for conduction in molecules can also be applied to derive the results of the previous chapters. We also identify the limitations of the approach by discussing a structurally similar, but much different device, the single electron transistor.

We are grateful to numerous colleagues who have been generous in sharing their insights and understanding with us. There are too many to thank individually, but one person stands out - our colleague, Supriyo Datta, whose simple and elegant understanding of nano-devices provided the inspiration for this monograph.

Chapter 1: Basic Concepts

1.1 Introduction

This chapter is a review of (or introduction to) some key concepts that will be needed as we examine nanotransistors. For the most part, concepts will be stated, not derived. A thorough introduction to these concepts can be found in Datta [1.1]. We also assume that the reader is acquainted with the basics of semiconductor physics (as discussed, for example, in [1.2]). For the quantum mechanical underpinning, see Datta [1.3] and for a more extensive discussion of semiclassical transport theory, see Lundstrom [1.4].

1.2 Distribution Functions

In equilibrium, the probability that a state at energy, E, is occupied is given by the Fermi function as

$$f_0 = \frac{1}{1 + e^{(E - E_F)/k_B T_L}},$$
\hfill (1.1)

where the subscript, 0, reminds us that the Fermi function is defined in equilibrium, and T_L is the lattice temperature. When states in the conduction band are located well above the Fermi level, the semiconductor is nondegenerate and eqn. (1.1) can be approximated as

$$f_0 \approx e^{(E_F-E)/k_B T_L} . \tag{1.2}$$

By writing the energy as the sum of potential and kinetic energies,

$$E = E_C + \frac{1}{2} m^* v^2 , \tag{1.3}$$

where E_C is the bottom of the conduction band, eqn. (1.2) can be written as

$$f_0 \approx e^{(E_F-E_C)/k_B T_L} e^{-m^* v^2/2k_B T_L} = C e^{-m^* v^2/2k_B T_L} , \tag{1.4}$$

where C is a constant. Equation (1.4) states that in a nondegenerate semiconductor, the carrier velocities are distributed in a Gaussian (or Maxwellian) distribution with the spread of the distribution related to the temperature of the carriers. Since $v^2 = v_x^2 + v_y^2 + v_z^2$, we can also write

$$f_0 \approx C' e^{-m^* v_x^2/2k_B T_L} , \tag{1.5}$$

which shows that the carrier velocities are distributed symmetrically about the x-axis (or for that matter, the y- and z-axes). The average velocity of the entire distribution is zero. Figure 1.1 is a sketch of a Maxwellian velocity distribution.

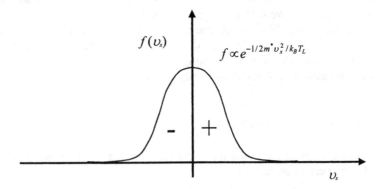

Figure 1.1   Illustration of a Maxwellian velocity distribution.   At equilibrium, the distribution is symmetric and centered at $v_x = 0$, so the average velocity is zero. The spread of the distribution is related to the temperature.

1.3 Three, Two, and One-Dimensional Carriers

Within an effective mass description, the wave function for electrons in a semiconductor is obtained by solving the Schrödinger equation,

$$-\frac{\hbar^2}{2m^*}\nabla^2\psi(\mathbf{r})+U(\mathbf{r})\psi(\mathbf{r})=E\psi(\mathbf{r}). \tag{1.6}$$

If the potential energy, $U(\mathbf{r})$, is constant, the solutions are plane waves

$$\psi(\mathbf{r})=\frac{1}{\sqrt{\Omega}}e^{i\mathbf{k}\cdot\mathbf{r}}, \tag{1.7a}$$

where Ω is a normalization volume.

Consider next electrons in a thin slab as shown in Fig. 1.2. These electrons are confined in the z-direction, but they are free to move in the x-y plane. The wave function of these quasi-two-dimensional electrons is found by solving eqn. (1.6) using separation of variables to find

$$\psi(\mathbf{r})=\phi(z)\,\psi(x,y)=\phi(z)\frac{1}{\sqrt{A}}e^{i(k_x x+k_y y)}=\phi(z)\frac{1}{\sqrt{A}}e^{i\mathbf{k}_\parallel\cdot\boldsymbol{\rho}}, \tag{1.7b}$$

where A is a normalization area and ρ is a vector in the x-y plane. If the confining potential is a simple, square well with infinite barriers, then $\phi(z)=\sqrt{2/W}\,\sin(k_n z)=\sqrt{2/W}\,\sin(n\pi z/W)$, where $n=1, 2, ...$

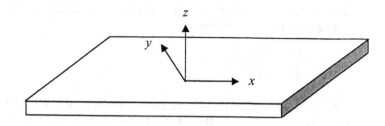

Figure 1.2   Quasi two-dimensional electrons that are confined in the z-direction but free to move in the x-y plane.

Consider next electrons in a thin wire as shown in Figure 1.3. These electrons are confined in the y- and z-directions, but they are free to move in the x direction. The wave function of these quasi-one-dimensional electrons is found solving eqn. (1.6) by separation of variables to find

$$\psi(\mathbf{r}) = \phi(y,z)\,\psi(x) = \phi(y,z)\frac{1}{\sqrt{L}}e^{ikx}. \tag{1.7c}$$

If the confining potential is a simple, square well with infinite barriers, then $\phi(y,z) = (2/W)\sin(m\pi y/W)\sin(n\pi z/W)$, where m, n = 1, 2, ...

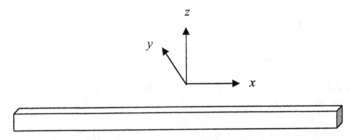

Figure 1.3   Quasi one-dimensional electrons that are confined in the y and z-directions but free to travel in the x-direction.

Quasi one- and two-dimensional electrons are produced by confining them so that they can move in one or two dimensions only. Confinement can be achieved electrostatically by producing a potential well with a gate potential, as for electrons in a bulk MOSFET, or by physically confining them to a thin, silicon film, as in a fully depleted silicon-on-insulator (SOI) MOSFET. Confinement leads to discrete energy levels. Consider the thin silicon slab shown in Fig. 1.2. If the confining potential is large, we may assume an infinite well to obtain

$$\varepsilon_n = \frac{\hbar^2 k_n^2}{2m^*} = \frac{\hbar^2 n^2 \pi^2}{2m^* W^2} \qquad n = 1,2,3... \tag{1.8}$$

If this *quantum well* represents the thin silicon body of a SOI MOSFET, the confining potential lies in the direction normal to the Si/SiO$_2$ interface and electrons are free to move in the x-y plane, the plane of the interface. Each energy level, therefore, represents a *subband* with many allowed k-states in the plane. The total energy is the sum of the energy due to confinement in the z-direction and the kinetic energy of motion in the x-y plane,

$$E(\mathbf{k}) = \varepsilon_n + \frac{\hbar^2 k_\parallel^2}{2m^*} = \qquad n = 1, 2, 3 \ldots \tag{1.9}$$

where $k_\parallel^2 = k_x^2 + k_y^2$. Because each subband has a density of states (Sec. 1.4), and given a Fermi level, we can compute the electron density in each subband (Sec. 1.5). Quantum confinement effectively raises the conduction band by an amount ε_n. In general, the shape of the potential well is more complicated than the simple box shown in Fig. 1.4, and the Schrödinger equation must be solved to determine the energy levels. The essential features remain, however. The subband energies increase and separate as the confinement increases (W decreases), light effective masses give high subband energies, and the carriers behave as quasi-two-dimensional carriers, free to move easily in only two dimensions. (Similarly, in a quantum wire, electrons are confined in two dimensions and behave as quasi-one-dimensional electrons.)

Figure 1.4b shows the constant energy surfaces for electrons in silicon, which are ellipsoids along the <100> directions. The ellipsoids are characterized by two effective masses, so the question of which effective mass to use in eqns. (1.8) and (1.9) arises. The subband energies are determined by the effective mass in the direction of the confining potential. Electrons in the two ellipsoids along the z-axis have a large effective mass (the longitudinal effective mass, $m_\ell = 0.98\, m_0$) and give rise to the unprimed series of subbands in Fig. 1.4a. On the other hand, electrons in the four other subbands have a light effective mass in the confinement direction (the transverse effective mass, $m_t = 0.19 m_0$) and produce the separate, primed series of subbands in Fig. 1.4c. So in eqn. (1.8), we use m_ℓ^* for the energy levels of the unprimed series and m_t^* for the primed series.

The heaviest effective mass gives the lowest subband energy, so the first level of the unprimed series is the lowest subband. For this level, the transverse effective mass in the plane, where electrons are free to move, is the light effective mass, m_t. This is the effective mass to use in eqn. (1.9). For the primed subbands, electrons have different effective masses when moving in the x- and y-directions, so an average effective mass must be used. When we evaluate the density-of-states, similar questions will arise. For electrons in silicon, it is often a good approximation to assume that all electrons occupy the lowest subband, especially when the confining potential is strong. Because it makes the bookkeeping easy, we will assume that only the lowest subband is occupied (see [1.5] for a more general treatment). For the first subband, the confinement energy is determined by the longitudinal effective mass and motion in the x-y plane by the transverse effective mass.

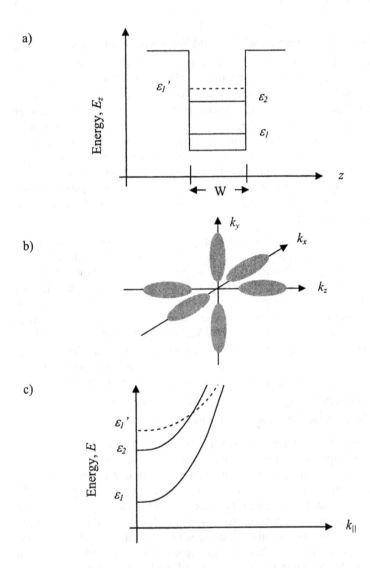

Figure 1.4   Quantum confinement in a well of width, W, with electrons free to move in the x-y plane.   (a) Energy levels for a Si quantum well.   (b) Constant energy surfaces for electrons in bulk Si, and (c) The $E(k)$ relation for electrons in the plane of the quantum well.

1.4 Density of States

For a bulk semiconductor with a volume, Ω, applying periodic boundary conditions to the wave function, eqn. (1.7a), leads to a discrete set of allowed k-states. The density of states in k-space is a constant,

$$N_{3D}(k)d^3k = 2 \times \frac{\Omega}{8\pi^3},$$

(1.10)

where Ω is a normalization volume, and the factor of 2 arises from spin degeneracy. We can use this result and the dispersion relation, $E(k)$, to derive the density of states in energy,

$$N_{3D}(E)dE = N_{3D}(k)d^3k.$$

(1.11)

For three-dimensional carriers, we find

$$D_{3D}(E) = \frac{N_{3D}(E)}{\Omega} = \frac{(2m^*)^{3/2}}{2\pi^2\hbar^3}\sqrt{E - E_C}.$$

(1.12)

Similar arguments can be used to derive the density of states for two-dimensional carriers. In k-space,

$$N_{2D}(k)d^2k = 2 \times \frac{A}{4\pi^2}d^2k,$$

(1.13)

where A is a normalization area in the two-dimensional plane. The two-dimensional density of states per unit energy and area is

$$D_{2D}(E) = \frac{m^*}{\pi\hbar^2}.$$

(1.14)

Finally, for one-dimensional carriers,

$$N_{1D}(k)dk = 2 \times \frac{L}{2\pi}dk,$$

(1.15)

where L is a normalization length for the wire. The one-dimensional density of states per unit energy and length is

$$D_{1D}(E) = \frac{\sqrt{2m^*}}{\pi\hbar}\frac{1}{\sqrt{E - E_C}}.$$

(1.16)

1.5   Carrier Densities

Recall that the equilibrium carrier density can be evaluated from the density of states and Fermi function as

$$n_o(E_F) = \int_0^\infty f_0(E, E_F) D(E) dE \qquad (1.17)$$

and expressed as a function of the Fermi energy. For three-dimensional carriers, the result is

$$n_o(E_F) = N_{3D} F_{1/2}\left[(E_F - E_C)/k_B T\right] = N_{3D} F_{1/2}(\eta_F) \quad cm^{-3}, \qquad (1.18)$$

where

$$N_{3D} = 2\left(\frac{2\pi m^* k_B T_L}{h^2}\right)^{3/2} \qquad (1.19)$$

is the *effective* density of states, $F_{1/2}$ is the Fermi-Dirac integral of order one-half [1.6], and $\eta_F = (E_F - E_C)/k_B T_L$. The Fermi-Dirac integral of order j is defined as [1.6]

$$F_j(\eta) \equiv \frac{1}{\Gamma(j+1)} \int_0^\infty \frac{\xi^j d\xi}{1 + e^{(\xi - \eta_F)}}. \qquad (1.20)$$

Under nondegenerate conditions, the Fermi-Dirac integral reduces to an exponential, and we find

$$n_o(E_F) = N_{3D}\, e^{(E_F - E_C)/k_B T_L}. \qquad (1.21)$$

Similarly, in two dimensions,

$$n_S(E_F) = N_{2D} \ln\left[1 + e^{(E_F - \varepsilon)/k_B T_L}\right] = N_{2D} F_0(\eta_F) \quad cm^{-2}, \qquad (1.22)$$

where ε is the bottom of the subband (only one is assumed) and

$$N_{2D} = \left(\frac{m^* k_B T_L}{\pi \hbar^2} \right) \tag{1.23}$$

is the two-dimensional effective density of states and F_0 is the Fermi-Dirac integral of order 0 [1.6]. Under nondegenerate conditions, eqn. (1.22) reduces to

$$n_S(E_F) = N_{2D}\, e^{(E_F - \varepsilon)/k_B T_L}. \tag{1.24}$$

Finally, for one dimensional carriers,

$$n_L(E_F) = N_{1D} F_{-1/2}(\eta_F) \quad \text{cm}^{-1}, \tag{1.25}$$

where

$$N_{1D} = \frac{\sqrt{2m^* k_B T_L / \pi}}{\hbar} \tag{1.26}$$

is the one-dimensional effective density of states, $F_{-1/2}(\eta_F)$ is the Fermi-Dirac integral of order $-1/2$ [1.6], and $\eta_F = (E_F - \varepsilon)/k_B T_L$. Under nondegenerate conditions, Eqn. (1.25) reduces to

$$n_L(E_F) = N_{1D}\, e^{(E_F - \varepsilon)/k_B T_L}, \tag{1.27}$$

where, again, ε is the bottom of the subband.

Equations (1.18), (1.22) and (1.25) relate the carrier densities to the Fermi level under arbitrary conditions. Under non-degenerate conditions, these expressions reduce to eqns. (1.21), (1.24), and (1.27). The general expressions can also be simplified for strongly degenerate conditions. Roughly speaking, carrier degeneracy occurs when the carrier density exceeds the effective density of states. At $T_L = 0K$, we have complete degeneracy; every state below E_F is occupied, and each state above E_F is empty. Using two-dimensional carriers as an example, eqn. (1.17) can be integrated to find

$$n_S(E_F) = (E_F - \varepsilon) \times D_{2D}. \quad (T_L = 0K) \tag{1.28}$$

Expressions for 3D and 1D carrier densities at $T_L = 0K$ can also be worked out.

In deriving expression for carrier densities, we worked in energy space beginning with eqn. (1.17). It is also possible to work in k-space. For example, for 2D carriers, the carrier density can be expressed as

$$n_S(k_F) = \frac{N_{2D}(k)}{A} \times \left(\pi k_F^2\right) = \frac{k_F^2}{2\pi}, \quad (T_L = 0K) \tag{1.29}$$

where the Fermi wave vector is obtained from

$$\left(E_F - \varepsilon\right) = \frac{\hbar^2 k_F^2}{2m^*} \tag{1.30}$$

1.6   Directed Moments

When evaluating carrier densities, we have summed the contributions from all k-states, but in a ballistic transistor, we will show that at a critical location, the positive and negative velocity k-states are populated according to different Fermi levels. We will, therefore, be interested in *directed moments* (integrals over only the positive k-states or only the negative k-states). Figure 1.5 shows what we mean by directed moments – still assuming equilibrium. Assuming 2D carriers at T = 0K, the carrier density located in positive velocity k-states is

$$n_S^+ = \left(E_F - \varepsilon\right) \times \frac{D_{2D}}{2} = \frac{k_F^2}{4\pi}, \quad (T_L = 0 \text{ K}) \tag{1.31}$$

and for $T > 0K$

$$n_S^+ = \frac{N_{2D}}{2} F_0\left[\left(E_F - \varepsilon\right)/k_B T_L\right], \tag{1.32}$$

where the factor of 2 occurs because only one-half of the k-states have positive velocity. Similar expressions apply to the negative-velocity carriers, and in equilibrium, where there is a single Fermi level, $n_S^+ = n_S^- = n_S/2$.

Similarly, we can evaluate the current carried by the positive k-states from the sum (assuming 2D carriers again)

$$J^+ = \frac{1}{A} \sum_{k_y, k_x > 0} q \upsilon_x f_0(E_k - E_F) \quad \text{A/cm} \tag{1.33}$$

which, at $T = 0$K gives

$$J^+ = \frac{q \hbar}{3\pi^2 m^*} k_F^3 = \frac{2q\sqrt{2m^*}}{3\pi^2 \hbar^2} (E_F - \varepsilon)^{3/2}. \quad (T_L = 0\text{K}) \tag{1.34a}$$

Alternatively, we can express the x-directed current density as

$$J^+ = q n^+ \left(\frac{4}{3\pi} \upsilon_F \right) = q n^+ \langle \upsilon^+ \rangle, \quad (T_L = 0 \text{ K}) \tag{1.34b}$$

where

$$\upsilon_F \equiv \frac{\hbar k_F}{m^*} \quad (T_L = 0\text{K}) \tag{1.35}$$

is the Fermi velocity. Note that the Fermi velocity is the velocity of carriers *at* the Fermi level, but that the average velocity of all carriers below the Fermi level is less than one-half of the Fermi velocity.

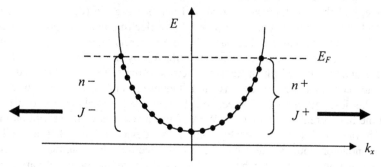

Figure 1.5   Illustration of the k-states at $T = 0$K and the definition of directed moments.

For $T_L > 0$K, eqn. (1.33) integrates to

$$J^+ = q n_S^+ \upsilon_T \left(\frac{F_{1/2}(\eta_{_F})}{F_0(\eta_{_F})} \right), \tag{1.36}$$

where

$$\upsilon_T \equiv \sqrt{\frac{2k_B T_L}{\pi m^*}}, \tag{1.37}$$

and the correction term for Fermi-Dirac statistics approaches one for a nondegenerate semiconductor. From eqn. (1.36), we see that the average x-directed velocity is

$$\bar{\upsilon}_T = \upsilon_T \frac{F_{1/2}(\eta_F)}{F_0(\eta_F)}. \tag{1.38}$$

Similar expressions apply to the negative velocity k-states, and in equilibrium $J^+ = J^-$ and $J = \left(J^+ - J^-\right) = 0$.

1.7 Ballistic Transport: Semiclassical

Consider the device sketched in Figure 1.6, which consists of a ballistic device attached to two contacts. We assume that strong scattering in the two contacts maintains thermal equilibrium there, but within the ballistic device, no scattering occurs. The source (contact 1) injects a thermal equilibrium flux into the device; some reflects from the potential barriers within the device, and the rest transmits across and enters the drain (contact 2). The contacts are taken to be perfect absorbers; any electron incident upon them is absorbed and thermalized. Similarly, the drain injects a thermal equilibrium flux into the device, some of which is reflected by the potential within the device and the rest of which transmits across to the source. In this section, we neglect quantum reflections and tunneling and treat electrons as semiclassical particles. Our objective is to compute the total electron density and the net current in the device. We will use a semiclassical description in which the local density-of-states within the device is just that of bulk semiconductor, but shifted by the local electrostatic potential. This approximation works well when the electrostatic potential does not vary too rapidly, so that quantum effects can be ignored.

To find how the k-states within the ballistic device are occupied, we solve the Boltzmann equation [1.4],

$$\frac{\partial f}{\partial t} + \upsilon_x \frac{\partial f}{\partial x} - qE_x \frac{\partial f}{\partial p_x} = \hat{C}f, \tag{1.39}$$

where $\hat{C}$ is the collision operator.

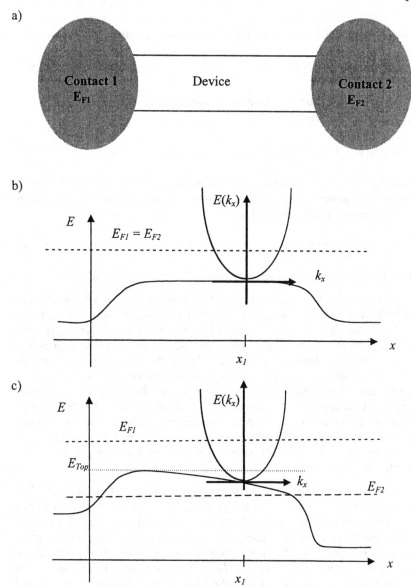

Figure 1.6   Sketch of a ballistic device with two contacts that function as reservoirs of thermal equilibrium carriers. (a) The device and the two contacts. (b) The energy band diagram under equilibrium conditions ($V_D = 0$). (c) The energy band diagram under bias ($V_D > 0$).

In equilibrium, detailed balance applies, so $\hat{C}f = 0$ and eqn. (1.39) becomes

$$v_x \frac{\partial f_0}{\partial x} - qE_x \frac{\partial f_0}{\partial p_x} = 0. \tag{1.40}$$

It may be shown that the solution to eqn. (1.40) is any function of the electron's total energy,

$$E = E_C(x) + E(\mathbf{p}) = E_C(x) + \frac{p^2}{2m^*}, \tag{1.41}$$

where $E_C(x)$ is the conduction band minimum versus position. We know from other arguments that under equilibrium conditions, the proper function of total energy is the Fermi function,

$$f_0(E) = \frac{1}{1 + e^{(E - E_F)/k_B T}}, \tag{1.42}$$

where the Fermi level, E_F, and temperature, T are constant in equilibrium.

Now consider the situation in Fig. 1.6c where a bias has been applied to the ballistic device. Although two thermal equilibrium fluxes are injected into the device, it is now very far from equilibrium. Since scattering is what drives the system towards equilibrium, the ballistic device under high bias is very far from equilibrium. Nevertheless, for the ballistic device, the relevant steady-state Boltzmann equation is still eqn. (1.40), the same equation as in equilibrium. The solution is again a function of the carrier's total kinetic energy. At the contacts, we know that the solution is a Fermi function, which specifies the functional dependence on energy. *For a ballistic device, therefore, the probability that a k-state is occupied is given by an equilibrium Fermi function.* The only difficulty is that we have two Fermi levels, so we need to decide which one to use.

Return again to Fig. 1.6c and consider how to fill the states at $x = x_1$. We know that the probability that a k-state is occupied is given by a Fermi function, so we only need to decide which Fermi level to use for each k-state. For the positive k-states with energy above E_{Top}, the top of the energy barrier, the states can only have been occupied by injection from the source, so the appropriate Fermi level to use is E_{F1}. Similarly, negative k-states with energy above E_{Top} can only be occupied by injection from the drain, so the appropriate Fermi level to use is E_{F2} Finally, for k-states below E_{Top}, both positive and negative velocity states are populated according to E_{F2}. (The negative velocity k-states are populated directly by injection from the drain and the positive k-states are populated when negative velocity carriers are reflected by the potential barrier.

Ballistic transport is a special kind of equilibrium. Each k-state is in equilibrium with the contact from which it was populated. Using this reasoning, one can compute the distribution function and any moment of it (e.g. carrier density, carrier velocity, etc.) at any location within the device. Figure 1.7 shows that the situation at the top of an energy barrier between two contacts is particularly simple – all positive k-states are filled according to the source Fermi level and all negative velocity states according to the drain Fermi level.

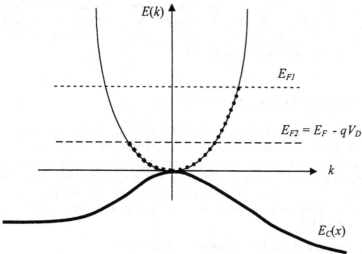

Figure 1.7   The $E(k)$ relation at the top of the barrier for the ballistic device of Fig. 1.6. Also shown are the source Fermi level, E_{FI}, and the drain Fermi level, $E_{F2} = E_{FI} - qV_D$. All positive k-states are occupied according to the source Fermi level and all negative k-states according to the drain Fermi level.

Using the results of Sec. 1.6, we can compute the two-dimensional carrier densities at the top of the barrier as

$$n_S^+ = \frac{N_{2D}}{2} F_0 \left[\left(E_F - \varepsilon \right) / k_B T_L \right]$$

(1.43a)

and

$$n_S^- = \frac{N_{2D}}{2} F_0 \left[\left(E_F - qV_D - \varepsilon \right) / k_B T_L \right].$$

(1.43b)

For the directed fluxes, we find

$$J^+ = n^+ \upsilon_T \left\{ \frac{F_{1/2}\left[(E_F - \varepsilon)/k_B T_L\right]}{F_0\left[(E_F - \varepsilon)/k_B T_L\right]} \right\} \tag{1.44a}$$

and

$$J^- = n^- \upsilon_T \left\{ \frac{F_{1/2}\left[(E_F - qV_D - \varepsilon)/k_B T_L\right]}{F_0\left[(E_F - qV_D - \varepsilon)/k_B T_L\right]} \right\}. \tag{1.44b}$$

We will use these ideas and expressions extensively when we develop the theory of the ballistic MOSFET in Chapter 3. Note that we have assumed 2D carriers, but similar expressions will be developed for 1D carriers in Chapter 5 when we discuss nanowire MOSFETs.

1.8   Ballistic Transport: Quantum

In Sec. 1.7, we described how to solve the ballistic Boltzmann equation for a device under bias. That semiclassical approach captures much of the physics of small devices, but new phenomena appear when quantum effects become important. In this section, we describe how the Schrödinger equation is solved for a ballistic device under bias. To keep the discussion simple, we assume a 1D quantum wire; extensions of this approach to 2D are described in [1.7].

We begin with the 1D Schrödinger equation,

$$-\frac{\hbar^2}{2m^*}\frac{\partial^2 \psi(x)}{\partial x^2} + E_s(x)\,\psi(x) = E\psi(x), \tag{1.45}$$

where $E_s(x)$ is the bottom of the first subband in the quantum wire (we assume a single subband that is uncoupled to higher subbands). Figure 1.8 is a sketch of the subband energy versus position for the first subband. As in Sec. 1.7, semi-infinite contacts are attached to the device at the source and drain ends. Because the potential in the contacts is assumed to be uniform, the solutions in the semi-infinite contacts are plane waves. If a unit amplitude wave is injected from the left (source) contact, then some portion reflects from the device and some transmits across and exits the perfectly absorbing right (drain) contact,

$$\psi(x) = 1 e^{ik_1 x} + r\, e^{-ik_1 x} \qquad\qquad x < 0 \tag{1.46a}$$

and

$$\psi(x) = t\, e^{ik_2 x}. \qquad\qquad x > L \tag{1.46b}$$

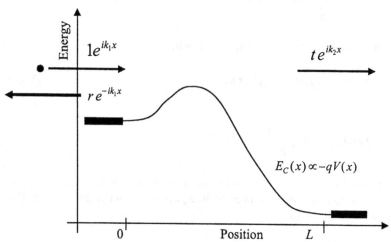

Figure 1.8   Illustration of a 1D device attached to two contacts in thermal equilibrium. Each contact launches electron waves into the device with an amplitude determined by the equilibrium Fermi function of the contact.

By solving eqn. (1.45) subject to the boundary conditions, eqns. (1.46a) and (1.46b), we find the wavefunction due to the injection of a unit amplitude wave from the source. To find the electron density vs. position within the device, we need to weight by the probability that state, k_1, with energy, $E(k_1)$, in the source is occupied and then sum over the distribution of electrons injected from the source,

$$n_1(x) = \frac{1}{L} \sum_{k_1} |\psi(x,k_1)|^2 f_0(E - E_F),\qquad(1.47)$$

where L is a normalization length in the contact.

Note that there is no well-defined $E(k)$ relation inside the device, because the wavefunctions are not plane waves; k_1 refers to the wavevector of an electron with energy, E, *in the source contact* and the subscript, 1, refers to injection from the first (left or source) contact. The probability that the state at energy, E, in the contact is occupied is given by a Fermi function because we assume that scattering maintains thermal equilibrium in the contacts.

By converting the sum over the injected k-states to an integral over energy, the electron density within the device can be expressed as

$$n_1(x) = \int_0^\infty n_1(x,E)dE, \qquad (1.48)$$

where we let the top of the band approach infinity and

$$n_1(x,E) = f_0(E - E_F)LDOS_1(x,E), \qquad (1.49)$$

where

$$LDOS_1(x,E) \equiv \left[\frac{1}{\pi} \frac{dk_1}{dE} |\psi(x,k_1)|^2 \right], \qquad (1.50)$$

is the *local density of states* (LDOS) due to injection from the source. For the total electron density, we need to integrate over energy and include the contribution of the drain,

$$n(x) = n_1(x) + n_2(x) =$$

$$\int_0^\infty \left[f_0(E - E_F)LDOS_1(x,E) + f_0(E - E_F + qV_D))LDOS_2(x,E) \right] dE \qquad (1.51)$$

where $LDOS_2$ is the local density of states due to injection from contact 2, the drain, which is computed from the wavefunction with boundary conditions analogous to eqns. (1.46a) and (1.46b).

To obtain the current, we evaluate

$$I_D = \frac{q}{L} \sum_{k_c} T_{1\to 2}(E)v_x(E_k) \left[f_0(E_k - E_F) - f_0(E_k - E_F + qV_D) \right], \qquad (1.52)$$

where $T_{1\to 2}(E) = T_{2\to 1}(E)$ is the current transmission coefficient from contact 1 to contact 2, and we define positive current as flowing into contact 2. After converting the sums to integrals, we find

$$I_D = \int_0^\infty I_D(E)dE, \qquad (1.53)$$

where

$$I_D(E) = \frac{2q}{h} T_{1\to 2}(E) \left[f_0(E - E_F) - f_0(E - E_F + qV_D) \right]. \qquad (1.54)$$

Finally, the current transmission coefficient $T_{1\to2}(E)$ is

$$T_{1\to2}(E) = \frac{I_{trans}}{I_{inc}} = 1 - \frac{I_{refl}}{I_{inc}} = 1 - |r|^2, \tag{1.55}$$

which, from eqn. (1.46a) can be expressed in terms of the known wavefunction as

$$T_{1\to2}(E) = 1 - |\psi(0) - 1|^2. \tag{1.56}$$

Numerical Solution
    To evaluate the expressions for the electron density and current, the wavefunction within the device must be known. To obtain it, we discretize eqn. (1.45) on a finite difference grid imposing the boundary conditions, eqns. (1.46a) and (1.46b), to find

$$[E\mathbf{I} - \mathbf{H} - \mathbf{\Sigma}_1 - \mathbf{\Sigma}_2]\mathbf{\psi} = i\mathbf{\gamma}_1, \tag{1.57}$$

where **H** is the N x N discretized Hamiltonian operator,

$$H_{i,j} = -t_0\delta_{i-1,j} + [2t_0 + E_s(x_i)]\delta_{i,j} - t_0\delta_{i+1,j}, \tag{1.58}$$

with an "on-site energy" of

$$t_0 = \frac{\hbar^2}{2m_x^* a^2}, \tag{1.59}$$

for a finite difference grid with a node spacing, a. The $N \times N$ *self-energy* matrices, $\mathbf{\Sigma}_1$ and $\mathbf{\Sigma}_2$, which account for the open boundary conditions are

$$\Sigma_1(i,j) = -t_0 e^{ik_1 a}\delta_{1,i}\delta_{1,j} \tag{1.60a}$$

$$\Sigma_2(i,j) = -t_0 e^{ik_2 a}\delta_{N,i}\delta_{N,j}. \tag{1.60b}$$

The N x 1 vector, $\mathbf{\gamma}_1$, is a source term that accounts for injection from contact 1 (the source). It has only one non-zero component, the first,

$$\gamma_1(1) = i\left[\Sigma_1(1,1) - \Sigma_1^*(1,1)\right] = 2t_0 \sin k_x a = \hbar\upsilon(k)/a. \tag{1.61}$$

For injection from the drain, we use boundary conditions analogous to eqns. (1.46a) and (1.46b), and the corresponding γ_2 vector has a non-zero component in position N.

It is important to note that our use of a discrete lattice alters the $E(k)$ in the contacts (where it is implicitly used as well). For a uniform contact with simple bands,

$$E(k_1) - E_c = \frac{\hbar^2 k_1^2}{2m^*},$$

(1.62a)

but the discrete grid modifies this relation to

$$E(k_1) - E_c = 2t_0 (1 - \cos k_1 a),$$

(1.62b)

where a is the grid spacing. The maximum energy must be well below the top of the band so that eqn. (1.62b) approximates eqn. (1.62a). (Note that the maximum allowed energy depends on the grid spacing, a.) For a given injection energy, E, k_1 and k_2 in eqns. (1.60) are found by solving eqn. (1.62b) in the source and drain contacts respectively. Finally, note that differentiation of eqn. (1.62b) with respect to $\hbar k$ gives $v(k)$ as used in eqn. (1.61).

The solution to eqn. (1.57) gives the wavefunction at each of the N finite difference nodes. The formal solution is

$$\psi = i\,G\gamma,$$

(1.63)

where

$$G = [E\mathbf{I} - \mathbf{H} - \mathbf{\Sigma}_1 - \mathbf{\Sigma}_2]^{-1}$$

(1.64)

is the retarded Green's function in a discrete basis. Because the matrix is tridiagonal and only one column of G is needed for computing Ψ, eqn. (1.57) can be efficiently solved; the carrier density and current are then evaluated from the computed wavefunctions. Alternatively, as discussed in the next section, one can express all of the results in terms of the N x N retarded Green's function [1.8].

When a quantum transport model is employed, quantum mechanical tunneling and reflections lead to differences from a semiclassical treatment. In the semiclassical approach, the local density of states within the device is just its value in the bulk [as given by eqn. (1.16)] shifted by the position dependent subband minimum, $E_C(x)$. In a quantum transport model, eqn.

(1.50) shows that there is an extra factor, $\psi^*(x)\psi(x)$ which accounts for tunneling and reflections. Carriers can tunnel under the potential barrier in a transistor (where the semiclassical model assumes a zero density of states) and they can be reflected even when they have an energy above the barrier where the semiclassical model assumes $T = 1$. Reference [1.7] shows several results that illustrate how quantum transport affects carrier transport in small MOSFETs.

1.9 The NEGF Formalism

Our conceptual view of a small device is summarized in Fig. 1.9. The device itself is connected to two contacts - large reservoirs where strong scattering maintains thermal equilibrium. Charged carriers may enter or exit the device from the two contacts. The strength of the coupling between the contacts and the device is described by *broadening functions*, Γ_1 and Γ_2. In equilibrium, $E_{FI} = E_{F2}$, but voltages applied to the contacts perturb equilibrium by raising or lowering the Fermi levels. The conductance of the device might also be modulated by a third electrode, a so-called gate. Because the charge within the device changes as a bias is applied, Poisson's equation must be solved to find the potential of the device. We are interested in computing the net charge within the device and the net current that flows between the two contacts self-consistently with Poisson's equation.

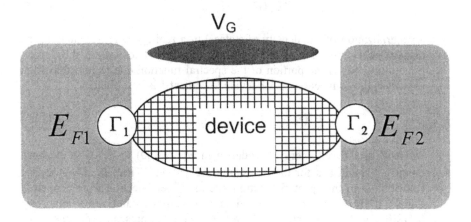

Figure 1.9   Schematic representation of a device.

Under ballistic conditions, the states within the device separate into two classes - those filled by injection from the source contact and those filled by injection from the drain. (States within the device that cannot be filled directly by injection from either contact may also exist. Such states can be filled indirectly by inelastic scattering processes discussed in the next section.) Populating the device with charge carriers consists, therefore, of two steps: i) computing the local density-of-states within the device fillable by each contact, and ii) filling the states according to the Fermi level of the appropriate contact. For 1D semiclassical ballistic transport, the local density of states fillable by a given contact is readily identified as shown in Fig. 1.6. For ballistic quantum transport, the local density-of-states is computed from the Schrödinger equation as was presented in eqn. (1.50).

The non-equilibrium Green's function (NEGF) formalism expresses the results of Sec. 1.8 in terms of $\mathbf{G}$, the retarded Green's function of the Schrödinger equation as given by eqn. (1.64) [1.8]. $\mathbf{G}$ is an N x N matrix obtained by discretizing the Hamiltonian using a finite difference (or finite element) discretization of an effective mass equation or by using a tight-binding approach with a set of molecular orbitals. (The size of the matrix, N, is the number of nodes in the finite difference grid or the total number of molecular orbitals.) After computing the Green's function, we evaluate the *spectral density*,

$$A_1(E) = \mathbf{G}\mathbf{\Gamma}_1\mathbf{G}^+, \tag{1.65}$$

a generalization of the density-of-states. In a real-space representation, the diagonal elements of $\mathbf{A}$ give the local density-of-states within the device. Equation (1.65) is the portion of the spectral function due to injection from contact 1; the connection of the device to contact 1 is described by

$$\mathbf{\Gamma}_1 = i\left(\mathbf{\Sigma}_1 - \mathbf{\Sigma}_1^+\right), \tag{1.66}$$

where $\mathbf{\Sigma}_1$ is the self-energy introduced in eqn. (1.60) to describe the open boundary conditions for the Schrödinger equation. There is also a second contribution to the spectral function due to carrier injection from contact 2. Figure 1.10a shows the computed local density-of-states for a nanoscale MOSFET under high gate and drain bias. Note the presence of states in the "forbidden region" below the bottom of the conduction band (more precisely, below the bottom of the first subband) and the strong quantum interference effects imposed on the slow variation with energy.

Having computed the spectral functions associated with the two contacts, we fill up the states according to the Fermi level of the appropriate contact to evaluate the *electron correlation function* due to injection from contact 1 as

$$G_1^n(E) = f_0(E_{F1} - E)\frac{\mathbf{A}_1}{2\pi},\tag{1.67}$$

where $f_0(E_{F1} - E)$ is the Fermi function of contact 1. Equation (1.67) is the portion of $\mathbf{G}^n$ due to injection from contact 1; there is a second contribution due to injection from contact 2. In a real space representation, the diagonal elements of $\mathbf{G}_n(E)$ give the energy-resolved electron density at each node, analogous to eqn. (1.49). The total electron density within the device is found by summing the contributions of the two contacts and integrating over energy as in eqn. (1.51). Figure 1.10b shows the energy-resolved electron density in a nanoscale MOSFET under high gate and drain bias. This result was obtained by filling up the local density-of-states shown in Fig. 1.10a according to the Fermi levels of the two contacts. The two distinct populations of carriers, one due to injection from the source and another due to injection from the drain, are clearly distinguished.

After evaluating the electron density within the device, Poisson's equation is solved to update the self-consistent electrostatic potential that appears in $\mathbf{H}$, and the process continues until it converges. Finally, under ballistic conditions, the current may be evaluated from the transmission coefficient at each energy [1.8]

$$T^{1-2}(E) = \text{trace}\left[\boldsymbol{\Gamma}_1 \mathbf{G} \boldsymbol{\Gamma}_2 \mathbf{G}^+\right].\tag{1.68}$$

Equation (1.68) expresses the transmission coefficient in terms of $\mathbf{G}$ rather than in terms of the wavefunction as in eqn. (1.56). The current is then obtained from eqns. (1.53) and (1.54).

As outlined in this section, the NEGF formalism is mathematically equivalent to solving the single particle Schrödinger equation with open boundary conditions. It provides, however, a powerful formalism that makes extensions to higher dimensions and molecular orbitals straightforward. More importantly, it has a solid theoretical foundation that permits us to go beyond the single particle Schrödinger equation when necessary. (See ref. [1.8] for a list of classic references.) One example is the treatment of inelastic scattering, as discussed in the next section.

a)

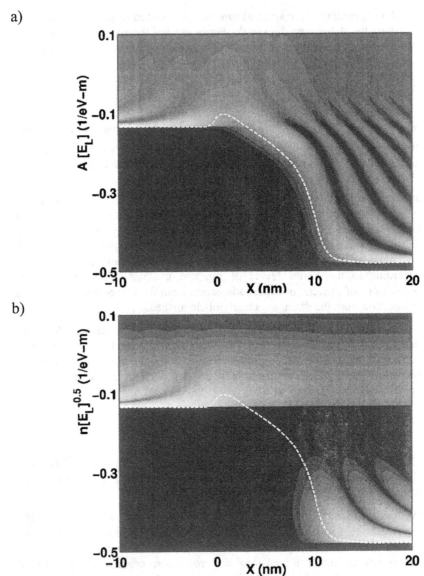

b)

Figure 1.10 Illustration of the local density of states and carrier density within a nanoscale
MOSFET. The dashed line is the conduction band edge vs. position. (a) The
local density of states vs. energy and position, and (b) the electron density vs.
energy and position. The light regions correspond to high densities. (See [1.7] for
a discussion of the device simulation techniques, reproduced with permission
from [1.9])

1.10 Scattering

Ballistic transport is easy to understand because there is a clear prescription for identifying how the states are populated by the two contacts. Scattering mixes the two populations of carriers, so that a state can be populated directly from one of the two contacts or indirectly by scattering out of another state.

To treat scattering within the NEGF formalism, we add a self-energy to the Green's function, so that eqn. (1.64) becomes

$$G = \left[E\mathbf{I} - \mathbf{H} - \Sigma_1 - \Sigma_2 - \Sigma_S \right]^{-1}, \tag{1.69}$$

where Σ_S is a self-energy that describes interactions with phonons and other scattering potentials. The self-energies, Σ_1 and Σ_2, account for the filling and emptying of states through the two contacts. Similarly, Σ_S accounts for the filling and emptying of states by in- and out-scattering. The self-energies in eqn. (1.69) shift and broaden the energy levels of the isolated device.

After evaluating the Green's function, the local density of states fillable by the two contacts and by scattering can be evaluated as in Eqn. (1.65). The states are then filled according to

$$G^n(E) = G\Sigma_1^{in}G^+ + G\Sigma_2^{in}G^+ + G\Sigma_S^{in}G^+, \tag{1.71}$$

where, from eqn. (1.67)

$$\Sigma_1^{in} = f_0\left(E_{F1} - E\right)i\left(\Sigma_1 - \Sigma_1^+\right), \tag{1.72a}$$

describes the filling of states from contact 1,

$$\Sigma_2^{in} = f_0\left(E_{F2} - E\right)i\left(\Sigma_2 - \Sigma_2^+\right) \tag{1.72b}$$

from contact 2, and

$$\Sigma_S^{in} = \left(N_\omega + \frac{1}{2} \pm \frac{1}{2}\right)D(\hbar\omega) \otimes G^n(E \pm \hbar\omega) \tag{1.72c}$$

the filling of states by in-scattering by phonons. The in-scattering function is proportional to the phonon occupation factor, the matrix, $D(\hbar\omega)$, which is related to the electron-phonon coupling, and electron correlation function at $E + \hbar\omega$ for phonon emission and at $E - \hbar\omega$ for phonon absorption. The symbol, $\otimes$, denotes element by element multiplication. For a fuller account of the treatment of scattering in the NEGF formalism, and references to the literature, see [1.1] and [1.8].

Figure 1.11, which should be compared to Fig. 1.10 for the ballistic case, illustrates the effect of inelastic scattering on a nanoscale device. Note first from Fig. 1.11a that the interference effects in the ballistic local density-of-states are washed out by the phase randomizing scattering. Also note from Fig. 1.11b that it is no longer possible to identify two distinct populations associated with injection from the two contacts.

1.11 Conventional Transport Theory

When collisions dominate, transport is commonly modeled by few low order moments of the Boltzmann equation, eqn. (1.39). A mathematical prescription for generating moment equations exists, but to formulate them in a tractable manner, numerous simplifying assumptions are required [1.4]. Moment equations provide a phenomenological description of transport that gives insight and quantitative results when properly calibrated. Our goal here is simply to acquaint the reader with some key concepts.

The zeroth moment of the Boltzmann equation gives the well-known continuity equation. In one dimension,

$$\frac{\partial n}{\partial t} = -\left(\frac{dF_n(x)}{dx}\right) + G_n - R_n \,, \tag{1.73}$$

where n is the electron density, F_n the electron flux, G_n the electron generation rate, and R_n the electron recombination rate. Equation (1.73) states that the electron density at a location increases with time if there is a net flux of electrons into the region (as described by the first term, minus the divergence of the electron flux). It also increases if electrons are being generated, but it decreases if electrons recombine. Any physical quantity must obey a conservation law like eqn. (1.73), so we may generalize eqn. (1.73) to [1.3, 1.4]

$$\frac{\partial n_\phi}{\partial t} = -\left(\frac{dF_\phi(x)}{dx}\right) + G_\phi - R_\phi \,, \tag{1.74}$$

where n_ϕ is the density of the physical quantity (a moment of the Boltzmann equation), F_ϕ is the flux associated with that quantity, G_ϕ, the rate at which the quantity is produced, and R_ϕ is the rate at which the quantity recombines.

a)

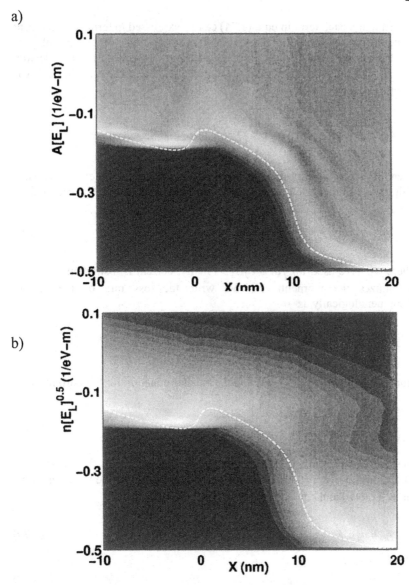

b)

Figure 1.11 Illustration of how scattering affects the local density of states and carrier density within a nanoscale MOSFET. The dashed line is the conduction band edge vs. position. (a) The local density of states vs. energy and position and b) the electron density vs. energy and position. The light regions correspond to high densities. (See [1.7] for a discussion of the device simulation techniques, reproduced with permission from [1.9])

Although each term in eqn. (1.74) can be expressed in terms of a moment of the distribution function [1.4], we will use it at a phenomenological level to establish the form of the transport equations. Consider first, the electron current. For the x-component of the current, $n_\phi = J_{nx}$. The associated flux is the "flux of a flux,"

$$F_x = -qn\upsilon_x\upsilon_x = -\frac{2q}{m^*}n\left(\frac{m^*\upsilon_x^2}{2}\right) = -\frac{2q}{m^*}W_{xx}, \qquad (1.75)$$

where W_{xx} is an energy-related tensor that is the kinetic energy associated with motion in the x-direction. Current is "generated" by the electric field,

$$G_\phi = -nq\frac{d\upsilon_x}{dt} = -\frac{nq}{m^*}\frac{dp_x}{dt} = \frac{nq^2}{m^*}E_x, \qquad (1.76)$$

where we have used $dp_x/dt = -qE_x$. Finally, current is lost when scattering randomizes the momentum. We write the loss rate of the current phenomenologically as

$$R_\phi = -\frac{J_{nx}}{\tau_m}. \qquad (1.77)$$

From eqns. (1.74) – (1.77), we find the balance equation for the current as

$$\frac{\partial J_{nx}}{\partial t} = \frac{2q}{m^*}\frac{dW_{xx}}{dx} + \frac{nq^2}{m^*}E_x - \frac{J_{nx}}{\tau_m}. \qquad (1.78)$$

The current typically changes slowly on the scale of the momentum relaxation time, τ_m, (typically a sub-picosecond time) so the time derivative can be ignored and eqn. (1.78) solved for

$$J_{nx} = nq\mu_n E_x + \frac{2}{3}\mu_n\frac{dW}{dx}, \qquad (1.79)$$

where

$$\mu_n = \frac{q\tau_m}{m^*} \qquad (1.80)$$

is the electron mobility and we have assumed equipartition of energy so that $W_{xx} = W/3$ where W is the total kinetic energy. Equation (1.80) is a drift-diffusion equation; electrons drift in electric fields and diffuse down kinetic energy gradients. Near equilibrium,

$$W = \frac{3}{2} n k_B T ,$$ (1.81)

so when T is uniform, eqn. (1.79) becomes

$$J_{nx} = n q \mu_n E_x + k_B T \mu_n \frac{dn}{dx} = n q \mu_n E_x + q D_n \frac{dn}{dx} ,$$ (1.82)

the familiar drift-diffusion equation. By using Eqn. (1.82) in the electron continuity equation, we get an equation that can be solved for the electron density within a device.

Since most devices contain regions with high electric fields, the assumption that $W = 3 n k_B T / 2$ is not a good one. The carrier energy enters directly into the second term of the transport equation, eqn. (1.79), but also enters indirectly because the mobility is energy dependent. (More energetic carriers scatter more frequently, so the mobility is lower for high energy carriers.) To treat transport more rigorously, we need an equation for the electron energy, which we get from another balance equation with $n_\phi = W$. In this case, the associated flux is an energy flux, J_{Wx}. The rate of increase of the electron energy is just the power input by the electric field, so

$$G_\phi = J_{nx} E_x .$$ (1.83)

Finally, we write the loss rate of kinetic energy due to scattering as

$$R_\phi = -\frac{(W - W_o)}{\tau_E} ,$$ (1.84)

where W_0 is the equilibrium energy density and τ_E the energy relaxation time. The energy balance equation becomes

$$\frac{\partial W}{\partial t} = -\frac{dJ_W}{dx} + J_{nx} E_x - \frac{(W - W_0)}{\tau_E} .$$ (1.85)

Note that the energy relaxation time is generally longer than the momentum relaxation time because phonon energies are small so that it takes several scattering events to thermalize an energetic carrier but only one to randomize its momentum.

To solve eqn. (1.85), we need to specify the energy current, which can be done by another balance equation. With appropriate simplifications, the energy flux becomes [1.4]

$$J_W = W \mu_E E_x + \frac{d(D_E W)}{dx} \tag{1.86}$$

where μ_E and D_E are the energy transport mobility and diffusion coefficient.

Equations (1.73), (1.79), (1.85) and (1.86) can now be solved self-consistently to simulate transport. Fig. 1.12 shows the result for bulk silicon with a constant electric field. At low electric fields, where $W \approx 3nk_B T / 2$, and we find that $\langle v_x \rangle \approx -\mu_n E_x$. For electric fields above $\approx 10^4$ V/cm, the kinetic energy increases, which increases the rate of scattering and lowers the mobility so that at high fields the velocity saturates at $\approx 10^7$ cm/s.

Electric fields well-above 10^4 V/cm are present in the channel of a nanoscale MOSFET. This is certainly high enough to cause velocity saturation in the bulk, but in a short, high field region, transients occur. Fig. 1.13 illustrates what happens. Electrons are injected and then begin to be accelerated by the electric field. Energy relaxation times are longer than momentum relaxation times, so the energy is slow to respond. The result is that the mobility is initially high, so the velocity can be very high. As the energy increases, however, scattering increases, the mobility drops, and the velocity eventually decreases to $\approx 10^7$ cm/s, the saturated velocity for electrons in bulk silicon. The spatial width of the transient is roughly 100nm. Since modern day channel length are below this value, strong velocity overshoot can be expected in modern devices.

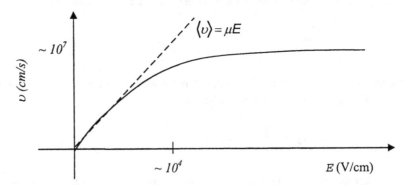

Figure 1.12   The average velocity vs. electric field for electrons in bulk silicon at room temperature.

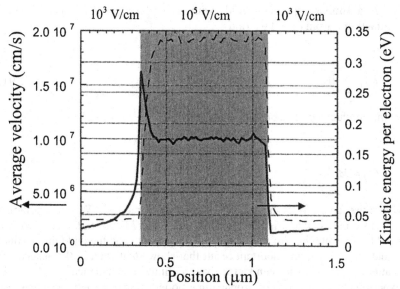

Figure 1.13 The average steady-state velocity vs. position for electrons injected into a short slab of silicon with a high electric field.

1.12 Resistance of a Ballistic Conductor at T = 0K

Consider the short, 1D ballistic conductor sketch in Fig. 1.14 for which we seek an expression for $I(V)$ at $T = 0$K. The current is the difference between the positive and negative currents,

$$I = I^+ - I^-$$ (1.87)

We evaluate I^+ from

$$I^+ = \frac{q}{L} \sum_{k>0} f(E_k) \upsilon_k = \frac{q}{\pi} \int_0^\infty f(E_k) \upsilon_k \, dk$$ (1.88)

At T = 0K, we find

$$I^+ = \frac{q}{\pi} \int_0^{k_F} \upsilon_k \, dk = \frac{q}{\pi} \int_0^{k_F} \frac{1}{\hbar} \frac{dE_k}{dk} \, dk$$ (1.89)

or

$$I^+ = \frac{2q}{h} \int_0^{E_F} dE = \frac{2q}{h} E_F.$$ (1.90)

For I^- a similar expression holds with E_F replaced by $E_F - qV$. Finally, subtracting I^- from (1.90) we find

$$I = \left(\frac{2q^2}{h}\right)V,$$ (1.91)

so the conductance of a ballistic conductor is

$$G_Q = M\left(\frac{2q^2}{h}\right),$$ (1.92)

where the factor, M, is the number of degenerate modes. (We have assumed that only one subband is occupied. If there are M subbands occupied, we need to compute the current for each one separately and then add the results.) Equation (1.92) is an important result that shows that even in the absence of scattering, a ballistic conductor has a finite conductance [1.1].   In experiments, the number of propagating modes, M, can sometimes be varied; the conductance is then quantized in units of $2q^2/h = 1/12.6\, K\Omega$.

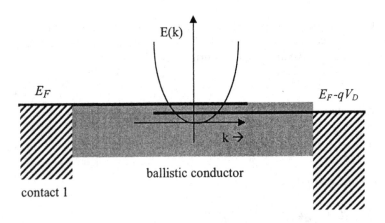

Figure 1.14   A short, one-dimensional ballistic conductor under small bias.

1.13 Coulomb Blockade

The discussion in Sec. 1.12 showed that a small ballistic conductor of the kind sketched in Fig. 1.6a displays an I-V characteristic of the type shown by the dashed line in Fig. 1.15. In devices of this type, the contacts are strongly coupled to the device. By "strongly coupled" we mean that the contacts are perfectly absorbing. When an electron in the device exits through a contact, it is completely absorbed; there is no amplitude for reflection at the contact, and an electron that enters the contact is thermalized before it can be re-injected according to the Fermi level of the contact. Consider, however, the device shown in Fig. 1.16, which consists of a device "island" weakly coupled to two leads that are connected to thermal equilibrium reservoirs. In this case, the I-V characteristics are much different, as shown in the solid line of Fig. 1.15. The current is not just reduced in magnitude by the weakly coupled contacts; no current at all flows until a voltage sufficient to overcome the *Coulomb blockade* is applied. This section is a very brief introduction to Coulomb blockade.

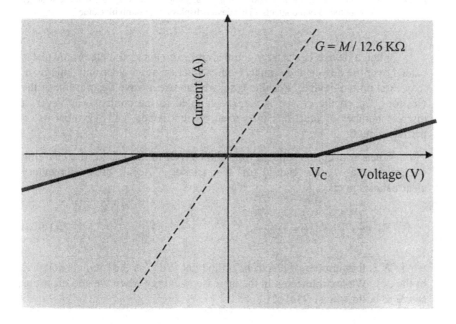

Figure 1.15 The current vs. voltage characteristics of two devices. The dashed line is for a small conductor strongly connected to the contacts, which displays a finite, Landauer conductance. The solid line is for a device weakly coupled to the two contacts (see Fig. 1.16), which displays so-called Coulomb blockade.

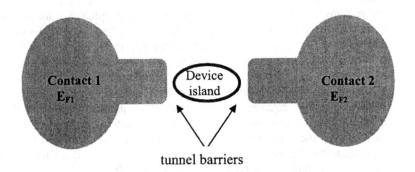

Figure 1.16   Sketch of a device "island" that is weakly coupled to two leads that are strongly coupled to two contacts. This device displays "Coulomb blockade."

When a device is strongly coupled to the contacts, the electronic states extend over the device and contacts, and a fraction of an electronic charge, q, may exist on the device. For this case, we can use a wave description of the electron. When the device is weakly coupled to the contacts, however, a discrete number of electrons may exist on the device, and we must use a particle picture.

When an electron tunnels from the contact to the island, the charging energy increases the energy of the island. Recall that elementary electrostatics gives

$$E_C = \frac{1}{2}CV^2 = \frac{Q^2}{2C} = \frac{N^2 q^2}{2C},$$ (1.93a)

where N is the number of electrons on the dot and C is the total capacitance of the dot. We are interested in the increase in energy when we add electrons, so we re-write eqn. (1.93a) as

$$E_C = \frac{N(N-1)q^2}{2C}$$ (1.93b)

to reflect the fact that it doesn't cost energy to add the first electron. (For large N, eqns. (1.93a) and (1.93b) are identical).

Equation (1.93b) is only part of the energy increase that occurs when an electron tunnels onto the island. To add charge to the island, the Fermi level must increase. Assuming two-dimensional carriers in an island of area, A, E_F is determined from

$$D_{2D}E_F(N)A = N. \tag{1.94}$$

Finally, note that the energy of electrons in the island may exceed that of electrons in the leads because of quantum confinement, which increases the energy by an amount, E_Q. The total energy as a function of the number of electrons, N, on the dot is

$$E(N) = \frac{q^2 N(N-1)}{2C} + E_F(N) + E_Q(N). \tag{1.95}$$

The change in energy to add one electron is

$$E(N+1) - E(N) = \frac{q^2}{C} + \Delta E, \tag{1.96}$$

where ΔE is the change in E_F and E_Q that occurs after an electron is added. Current cannot flow in the device of Fig. 1.16 unless this energy is available – a phenomenon known as Coulomb blockade.

One source of energy is thermal energy; if the thermal energy is large enough, then Coulomb blockade is not observed. To observe Coulomb blockade, we require that

$$\frac{q^2}{C} \gg k_B T_L. \tag{1.97}$$

If eqn. (1.97) holds, then current cannot flow until the drain bias exceeds $V_C \approx q/C$ (assuming that ΔE is small), which explains the *I-V* characteristic of Fig. 1.15.

To observe Coulomb blockade at room temperature, we require that $C \ll 3$ aF (1 aF $= 10^{-18}$F), an extremely small value. If we estimate the diameter, d, of the island required from the formula for the self capacitance of a sphere, $C = 2\pi\varepsilon d$, we find that $d < 1$nm. Producing an island small enough to observe Coulomb blockade at room temperature is a challenge when done by lithography, but the required size is that of a typical small molecule.

Two conditions must be met to observe Coulomb blockade: i) the charging energy must exceed the thermal energy, and ii) the contacts must be weakly coupled to the device. The second condition ensures that the eigenstates of the island are spatially localized on the island so that a discrete number of electrons may be placed on the island. Weak coupling is ensured by requiring that the resistance of the tunneling contact, R_T, exceeds a critical value that we can estimate from the Uncertainty Principle [1.10]

$$\Delta E \, \Delta t > \hbar/2 \,. \tag{1.98}$$

The characteristic time to move charge on and off of the island is

$$\Delta t \sim R_T C \,. \tag{1.99}$$

Single electron charging introduces an uncertainty in energy of

$$\Delta E = \frac{q^2}{C} \,. \tag{1.100}$$

By inserting eqns. (1.99) and (1.100) into (1.98), we find

$$R_T \gg \frac{\hbar}{2q^2} \text{ or } R_T \gg R_Q \tag{1.101}$$

Equation (1.101) is the answer that we might have expected. For strongly coupled contacts, eqn. (1.92) gives a resistance of $R_Q = h/2q^2 M$. For weak coupling, the resistance of the tunneling contacts should be much greater than this value. For a tunneling contact, eqn. (1.92) is modified by multiplying by a current transmission coefficient, T (where $0 \le T \le 1$), so

$$R_T = \frac{1}{MT}\left(\frac{h}{2q^2}\right). \qquad\qquad (1.102)$$

From (1.102) and (1.101), we conclude that when $T \ll 1$, the contacts are weakly coupled, a particle picture should be used, and single electron charging effects can occur. When $T \approx 1$, the wave picture should be used, and single electron charging does not occur.

A number of interesting effects related to Coulomb blockade can occur in practice. For example, background charge can shift or even remove V_C, asymmetric contacts can produce a staircase characteristic, and by gating the island, single electron transistors can be produced. These topics will be discussed in Chapter 6 (see also ref. [1.10]). Finally, note that if one were to apply the NEGF formalism as presented in Sec. 1.9 to this problem by reducing Γ to reduce the coupling to the contacts, one would NOT obtain the correct result with an integral number of electrons on the island. Coulomb blockade is actually an example of the breakdown of the single particle picture that the formalism of Sec. 1.9 is based on [1.8].

1.14 Summary

Several key concepts that will be used in later chapters have been reviewed. The concept of carrier confinement, the relation of 1, 2, and 3D carrier densities to the Fermi level, and the notion of directed moments are especially important. Understanding how the eigenstates are populated in a ballistic device and appreciating the difference between semiclassical and quantum treatments are also important. Finally, the idea of a quantum conductance, $2q^2/h$, will play an important role in our development of models for nanotransistors, for which a wave picture can be used. Understanding when to use a wave approach and when a particle picture is needed is important for interpreting experiments and for exploring new devices.

Chapter 1 References

[1.1]   S. Datta, *Electronic Conduction in Mesoscopic Systems*, Cambridge University Press, Cambridge, UK, 1996.

[1.2]   R. F. Pierret, *Advanced Semiconductor Fundamentals*, Addison-Wesley, Reading, Massachusetts, 1987.

[1.3]   S. Datta, *Quantum Phenomena*, Addison-Wesley, 1989.

[1.4]   M. S. Lundstrom, *Fundamentals of Carrier Transport*, 2nd Ed., Cambridge University Press, Cambridge, UK, 2000.

[1.5]   F. Assad, Z. Ren, D. Vasileska, S. Datta, and M. S. Lundstrom, "On the Performance Limits of Silicon MOSFETs: A Theoretical Study," *IEEE Trans. Electron Dev.*, **47**, pp. 232-240, 2000.

[1.6]   J. S. Blakemore, "Approximations for Fermi-Dirac Integrals, Especially the Functions $F_{1/2}(\eta)$ to Describe Electron Density in a Semiconductor," *Solid-State Electron.*, **25**, pp. 1067, 1982.

[1.7]   Z.Ren, R.Venugopal, S.Goasguen, S.Datta, and M.S. Lundstrom "nanoMOS 2.5: A Two -Dimensional Simulator for Quantum Transport in Double-Gate MOSFETs," *IEEE Trans. Electron. Dev.*, special issue on Nanoelectronics, **50**, pp. 1914-1925, 2003.

[1.8]   S. Datta, *Quantum Transport Atom to Transistor*, Cambridge University Press, Cambridge, UK. 2005.

[1.9]   R. Venugopal, M. Paulsson, S. Goasguen, S. Datta, and M. Lundstrom, " A Simple Quantum Mechanical Treatment of Scattering in Nanosacle Transistors," *J. of Appl. Phys.*, **93**, pp. 5613-5625, 2003.

[1.10] K. Likharev, "Electronics Below 10 nm," in: *Nano and Giga Challenges in Microelectronics*, ed. by J. Greer et al., Elsevier, Amsterdam, pp. 27-68, 2003.

Chapter 2: Devices, Circuits, and Systems

2.1   Introduction

The integrated circuit made modern day information processing and communications systems possible. Its basic functional element is the transistor, most commonly a silicon metal oxide semiconductor field-effect transistor (MOSFET). For the past forty years, MOSFET scaling (the reduction of its critical dimension by a factor of about $\sqrt{2}$ each technology generation, approximately 18 months) has driven Moore's Law (the doubling of the number of transistors per integrated circuit each technology generation). It now appears that the silicon MOSFET will reach its scaling limit within a decade or so [2.1, 2.2], and devices to complement or replace the silicon MOSFET are being explored [2.3].

This chapter is a brief overview of MOSFET essentials and a quick introduction to the bipolar transistor. The chapter provides some context for exploring new devices and an opportunity to discuss three important points. First, we discuss charge control by a gate electrode, which modulates the transistor's current. Electrostatics is likely to be similarly important for transistors that follow the MOSFET. Second, we discuss the characteristics of devices that make them useful in high-density, high-speed digital systems. Finally, we examine the fundamental limits that apply to any electronic switching device used for conventional, digital logic.

2.2   The MOSFET

Figure 2.1 illustrates the physical structure of two different kinds of MOSFETs. An n-channel bulk MOSFET is built on a p-type substrate with deep n^+ regions to facilitate contact to the source and drain. Shallow n^+ junctions connect the source and drain to the p-type channel. A thin gate oxide (typically still SiO_2 and about 1-2nm thick) separates the silicon channel from the gate electrode. Figure 2.1b shows a double gate MOSFET, which is built on a thin silicon film and with gates above and below the channel [2.2]. Numerous variations exist. Examples include the FinFET, a type of double gate MOSFET [2.3, 2.4], the tri-gate transistor [2.5], and the gate-all-around MOSFET [2.6].

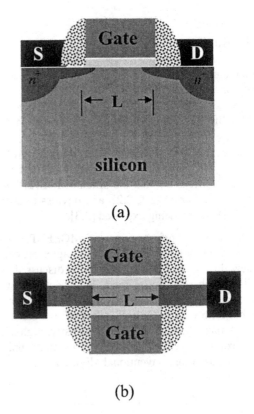

(a)

(b)

Fig. 2.1   Cross sectional sketches of: a) a bulk, silicon MOSFET and b) a double-gate MOSFET. The third dimension, the width, *W*, of the MOSFET, is into the page.

For any type of MOSFET, the gate voltage modulates the conductivity of the p-type channel by raising or lowering the height of an energy barrier between the source and channel, as shown in Fig. 2.2. Under low drain voltages (Fig. 2.2a), the device operates like a resistor with the gate voltage controlling the resistance, while under high drain bias (Fig. 2.2b), the device operates like a current source with the gate voltage controlling the magnitude of the current. The transistor designer's challenge is to engineer an appropriate energy barrier between the source and drain so that the device can be turned off while at the same time designing a gate structure that can effectively modulate the barrier and turn the transistor on. The design of a bulk MOSFET for proper electrical performance involves producing sophisticated two-dimensional doping profiles in the p-type channel, an ultra-thin gate oxide, and heavily doped, ultra-shallow source/drain extensions [2.7]. Double gate, tri-gate, and gate-all-around MOSFETs provides strong gate control of the channel conductivity, which allows the source and drain to be placed more closely. The channel length, L, sets the scale of the device. Device scaling refers to the process of shrinking L to reduce the device size, but a complete MOSFET is typically 10-15 times larger than L. The associated dimensions (oxide thickness, shallow extension junction depth, etc.) also need to be reduced accordingly to maintain good electrical characteristics.

Figure 2.3 sketches the drain current, I_D, vs. drain-to-source voltage, V_{DS}, characteristics of the MOSFET. Because the MOSFET has four terminals there are several ways to plot these characteristics. In Fig. 2.3a, we plot I_D vs. V_{GS} on both linear and logarithmic axes. On a linear scale, essentially no current flows until the gate voltage reaches a critical value, the threshold voltage, V_T. On a logarithmic scale, we see that the drain current actually increases exponentially for $0 < V_{GS} < V_T$. Above threshold, I_D varies as the gate overdrive, $(V_{GS} - V_T)$ to a characteristic power, α. For low V_{DS}, $\alpha = 1$, but for high V_{DS}, $1 \le \alpha \le 2$. The maximum current, known as the on-current, occurs when the power supply voltage is applied between the drain and source and between the gate and source.

Figure 2.3b plots I_D vs. V_{DS} with V_{GS} as a parameter. For low V_{DS}, the MOSFET operates like a gate voltage dependent resistor, but for high V_{DS}, it operates more like a gate voltage controlled current source (with a finite output conductance). The voltage that separates these two regions is the so-called "drain saturation voltage," V_{Dsat}.

Figure 2.3c plots log I_D vs. V_{GS} at low and high drain voltages. The subthreshold region is characterized by its slope or, equivalently, the subthreshold swing, S, which is the number of millivolts of increase in gate voltage needed to increase the drain current by a factor of 10. For well-designed MOSFETs, $S < 80$ mV/decade; the theoretical lower limit is 60 mV/decade at room temperature. Another important performance metric is the *off-current*, the current that flows when the $V_{DS} = V_{DD}$ and $V_{GS} = 0$. A good transistor should display a high on-current, a low off-current, and a rapid transition between the off and on states (i.e. a small S).

(a)

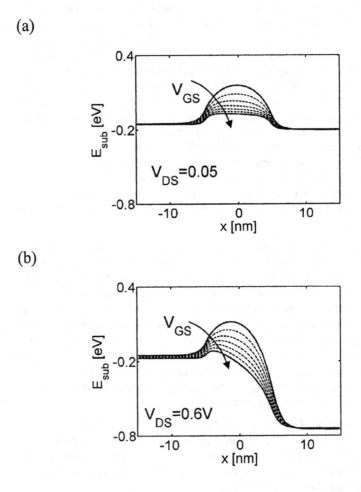

(b)

Fig. 2.2   Sketch of the minimum electron energy vs. position showing how an increasing gate voltage lowers the energy barrier between the source and drain. Two cases are shown: a) low drain voltage and b) high drain voltage.

(a)

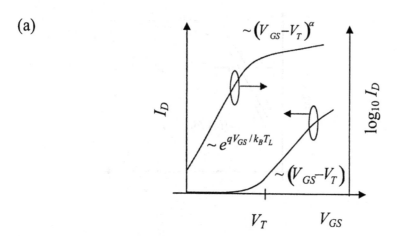

(b)

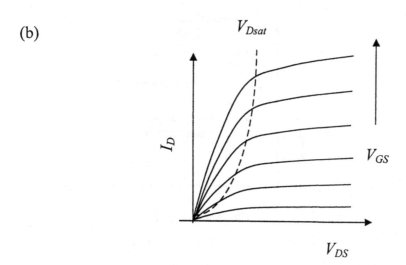

Fig. 2.3   The current vs. voltage characteristics of a MOSFET. a) I_D vs. V_{GS} for a fixed V_{DS}. b) I_D vs. V_{DS} with V_{GS} as a parameter.

(c)

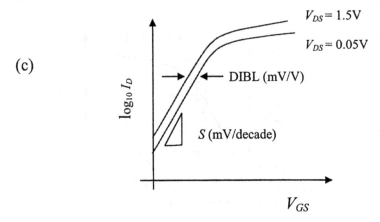

(d)

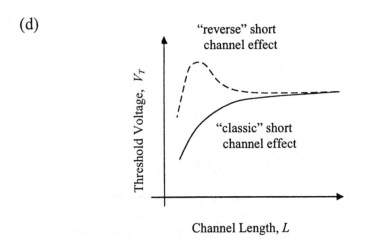

Fig. 2.3   The current vs. voltage characteristics of a MOSFET. c) log I_D vs. V_{GS} at low
               and high drain bias. d) The threshold voltage, V_T, vs. channel length.

Fig. 2.3c also shows that the I_D-V_{GS} characteristics for low and high V_{DS} are translated horizontally (for poorly designed MOSFETs, S also changes). The translation is known as DIBL (drain-induced barrier lowering) and is characterized by the number of millivolts of translation per volt of change in drain voltage. Well-designed MOSFETs typically have DIBL < 100 mV/V.

Finally, Fig. 2.3d sketches V_T vs. channel length, L. Two-dimensional electrostatic effects tend to reduce V_T as L decreases. Laterally non-uniform doping profiles can reverse this effect and produce an initial increase in V_T as L decreases. The goal of the transistor designer is to make V_T as nearly independent of L as possible.

In this section, we have described the physical structure and terminal I-V characteristic of the MOSFET. In the following sections, we highlight a few important concepts that we will make use of in later chapters. For a more extensive treatment, refer to Taur and Ning [2.7].

2.3   1D MOS Electrostatics

The most important thing to understand about a MOSFET is MOS electrostatics. We begin in 1D in equilibrium by examining a long channel MOSFET with $V_{DS} = 0$ (Fig. 2.4). Near the middle of the channel, there is no variation of potential with x. A positive voltage on the gate lowers the electron energy and bends the bands down by an amount, $q\psi_s$, as illustrated in Fig. 2.4b and 2.4c. Our goal is to determine how the charge in the semiconductor, Q_S in C/cm^2 varies with surface potential, ψ_s , or alternatively, with gate voltage, V_{GS}. Later, we will seek to understand two-dimensional electrostatics and the influence of V_{DS}.

A direct approach to finding $Q_s(\psi_s)$ is to solve Poisson's equation,

$$\frac{d^2\psi}{dy^2} = -\frac{\rho}{\varepsilon_{Si}} . \tag{2.1}$$

The charge density, ρ, is related to the mobile carrier densities, $n(y)$ and $p(y)$, which are related to band bending and, therefore, to $\psi(y)$. The result is the so-called "Poisson-Boltzmann equation," which can be solved numerically for $Q_S(\psi_S)$ [2.7, 2.8]. We seek a simpler approach in order to understand the essence of the problem.

(a)

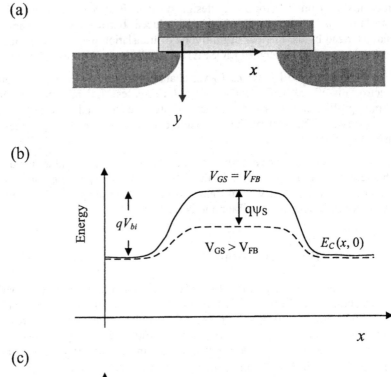

(b)

(c)

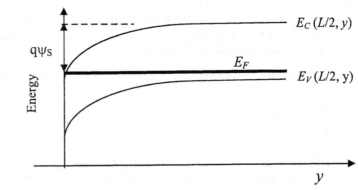

Fig. 2.4   Energy band diagrams for a silicon MOSFET in equilibrium. (a) the coordinate
system, (b) $E_C(x,0)$, the conduction band edge at the oxide/silicon interface
along the channel from source to drain. (c) $E_C(L/2,y)$, the conduction band
edge in the middle of the channel vs. position into the silicon bulk.

The charge in the semiconductor has two components,

$$Q_S(\psi_S) = Q_B(\psi_S) + Q_i(\psi_S), \tag{2.2}$$

where Q_B, the bulk charge, is due to the depletion of majority carriers, and Q_i, the mobile charge, is due to inversion (or accumulation) layers of mobile carriers. Consider first a small, positive ψ_S for which we have only bulk charge described by depletion layer theory as [2.7, 2.8]

$$Q_B(\psi_S) = -qN_A W_D = -\sqrt{2q\varepsilon_{Si} N_A \psi_S}, \tag{2.3}$$

where W_D is the width of the surface depletion layer. The small mobile charge, $Q_i = -qn_S$ C/cm^2, does not affect the electrostatics, but it does give rise to the subthreshold current. Note that

$$n(y) = n_{op} e^{q\psi(y)/k_B T_L} = \frac{n_i^2}{N_A} e^{q\psi(y)/k_B T_L} \text{ cm}^{-3}, \tag{2.4}$$

where n_{op} is the equilibrium minority electron density in the p-type bulk. The integrated electron density per cm^2 is

$$n_S = \int_0^\infty n(y)dy = \int_0^\infty \frac{n(y)}{d\psi/dy} d\psi \text{ cm}^{-2}. \tag{2.5}$$

Since $n(y)$ falls rapidly for $y > 0$, we can approximate eq. (2.5) as

$$n_S(\psi_S) \cong -\frac{1}{E_S} \int_{\psi_S}^0 \frac{n_i^2}{N_A} e^{q\psi/k_B T_L} d\psi = \left(\frac{k_B T_L / q}{E_S}\right) \frac{n_i^2}{N_A} e^{q\psi_S/k_B T_L}, \tag{2.6}$$

where E_S is the electric field at the surface of the silicon. Finally, we have

$$Q_i(\psi_S) = -q\, n_S(\psi_S) = -q \left(\frac{k_B T_L / q}{E_S}\right) n(0) \equiv -q\, n(0) W_{inv}, \tag{2.7}$$

where $n(0)$ is the electron concentration per cm^3 at $y = 0$, and $(k_B T_L / q)/E_S$ is interpreted as the effective width of the inversion layer, W_{inv}.

As long as ψ_S is not too large, $Q_i(\psi_S) << Q_B(\psi_S)$ and

$$Q_S(\psi_S) \approx Q_B(\psi_S) \propto \sqrt{\psi_S} \qquad\qquad \psi_S < 2\psi_B \tag{2.8}$$

as illustrated in Fig. 2.5a. When ψ_S is greater than about

$$2\psi_B = \frac{2k_BT_L}{q}\ln(N_A/n_i),$$ (2.9)

then $Q_i(\psi_S) \gg Q_B(\psi_S)$. In this case,

$$Q_S \approx Q_i \approx -\varepsilon_{Si}E_S,$$ (2.10)

which can be used in eq. (2.7) to find

$$Q_S \propto e^{q\psi_S/2k_BT_L} \quad . \qquad\qquad\qquad \psi_S > 2\psi_B \qquad (2.11)$$

Similar arguments apply to the heavily accumulated region, $\psi_S < 0$, where the charge is due to the accumulation of majority carrier holes, so putting it all together, we obtain the $Q_S(\psi_S)$ characteristic as sketched in Fig. 2.5a. Our simple arguments establish the shape of the $Q_S(\psi_S)$ characteristic in accumulation ($\psi_S < 0$), depletion ($0 < \psi_S < 2\psi_B$), and inversion ($\psi_S > 2\psi_B$). The complete $Q_S(\psi_S)$ can be evaluated by solving the Poisson-Boltzmann equation [2.7, 2.8].

The Semiconductor Charge vs. Gate Voltage:

Having understood the $Q_S(\psi_S)$ characteristic, we now turn to $Q_S(V_{GS})$. The voltage at the gate is [2.7, 2.8]

$$V'_{GS} = \psi_S + \Delta V_{ox} = \psi_S - \frac{Q_S(\psi_S)}{C_{ox}},$$ (2.12a)

where

$$C_{ox} \equiv \frac{\varepsilon_{ox}}{t_{ox}},$$ (2.12b)

and the second expression arises because $\Delta V_{ox} = E_{ox}t_{ox}$ and $\varepsilon_{ox}E_{ox} = -Q_S$. In Eqn. (2.12a), $V'_{GS} = V_{GS} - V_{FB}$ where V_{FB} is the flatband voltage, the voltage at which there is no band bending in the semiconductor. Its value is determined by the gate to semiconductor workfunction difference and by charges at the oxide-silicon interface [2.7, 2.8]. So, having determined $Q_S(\psi_S)$, we can use (2.12) to translate it to a $Q_S(V'_{GS})$ characteristic. The resulting $Q_S(V'_{GS})$ characteristic sketched in Fig. 2.5b should be compared to the $Q_S(\psi_S)$ characteristic of Fig. 2.5a.

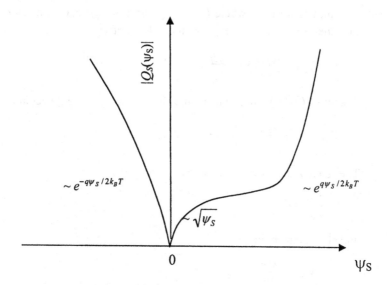

Fig. 2.5a   Charge in the semiconductor, Q_S, vs. surface potential for p-type silicon.

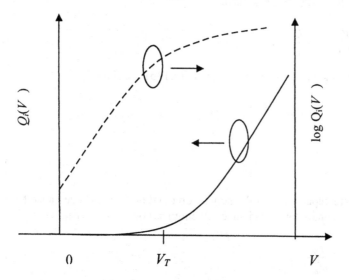

Fig. 2.5b   Inversion layer charge vs. gate voltage. On a linear scale (solid line), we see that the inversion layer charge becomes significant above the threshold voltage, V_T where it varies linearly with $(V_G - V_T)$. On a logarithmic scale (dashed line), we see that the inversion layer charge varies exponentially with gate voltage below threshold.

Above threshold, where $Q_i \gg Q_B$, the $Q_i(V_{GS})$ relation becomes simple. Consider a Taylor series expansion of Q_i about V_T

$$Q_i(V_{GS}) = Q_i(V_T) + \frac{dQ_i}{dV_{GS}}(V_{GS} - V_T). \tag{2.13a}$$

Assuming $Q_i(V_T) \cong 0$ and using the chain rule, eq. (2.13a) becomes

$$Q_i(V_{GS}) = \frac{dQ_i}{d\psi_S} \cdot \frac{d\psi_S}{dV_{GS}}(V_{GS} - V_T). \tag{2.13b}$$

The *semiconductor capacitance* is

$$-\frac{dQ_S}{d\psi_S} = C_S, \tag{2.14a}$$

and above threshold,

$$C_S \approx -\frac{dQ_i}{d\psi_S} = C_{inv}. \tag{2.14b}$$

From eq. (2.12a), we also have

$$\frac{dV_{GS}}{d\psi_S} = 1 + \frac{C_{inv}}{C_{ox}}. \tag{2.15}$$

After using eqs. (2.14b) and (2.15) in (2.13b), we find

$$Q_i \cong -C_G(V_{GS} - V_T), \qquad V_{GS} > V_T \tag{2.16a}$$

where

$$C_G \equiv \frac{C_{ox}C_{inv}}{C_{ox} + C_{inv}} < C_{ox}. \tag{2.16b}$$

The gate capacitance is the series combination of the oxide capacitance and the semiconductor capacitance. Phenomenologically, we can write

$$C_S = \frac{\varepsilon_S}{W_{inv}}, \tag{2.17}$$

which provides another way to define the inversion layer width. (See eqn. (2.7) for the other definition.) Because inversion layers are typically thin (~ 5 nm), MOSFET analysis has traditionally assumed that $C_{inv} \gg C_{ox}$ so that $C_G \approx C_{ox}$ for above threshold operation.

An aside on the quantum capacitance:

The inversion layer capacitance, is closely related to the "quantum capacitance," and is becoming increasingly important as device scaling decreases t_{ox} and increases C_{ox}. We can evaluate C_{inv} from eqns. (2.7) and (2.14), but (2.7) assumes Boltzmann statistics, and above threshold, degenerate statistics should be used. As an illustration, consider a fully degenerate ($T = 0K$) case for a quantum well with one subband occupied. As shown in Fig. 2.6, a gate potential raises or lowers the subband. The quantum capacitance is

$$C_S = q \frac{\partial n_S}{\partial \psi_S}, \tag{2.18}$$

and

$$n_s = D_{2D} \times (E_F - \varepsilon_1) \tag{2.19}$$

where D_{2D} is the constant 2D density-of-states and

$$\varepsilon_1 = \varepsilon_{10} - q\psi_S. \tag{2.20}$$

Using eqs. (2.19) and (2.20) in (2.18) we find

$$C_S = q^2 D_{2D}, \qquad (T_L = 0K) \tag{2.21}$$

so the quantum capacitance is proportional to the density of states. According to eqn. (2.16), a large inversion layer capacitance is beneficial for inducing charge in a semiconductor. From this perspective, a large effective mass is beneficial, but as we shall see later, transport suffers when the effective mass is large.

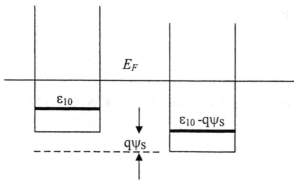

Fig. 2.6   Illustration of the origin of quantum capacitance.

Equation (2.21) shows that the $T_L = 0K$ quantum capacitance is proportional to the density of states. To derive the quantum capacitance for $T_L > 0K$, we generalize eqn. (2.19) to

$$n_S = \int_{\varepsilon_1}^{\infty} D_{2D} f(E - E_F) dE = \int_{0}^{\infty} D_{2D} f(E' - E_F + \varepsilon_{10} - q\psi_S) dE' \qquad (2.22)$$

from which we can derive the quantum capacitance as in eqn. (2.18) as

$$C_S = q \frac{\partial n_S}{\partial \psi_S} = q \int_{0}^{\infty} D_{2D} \frac{\partial f}{\partial \psi_S} dE = q^2 \int_{0}^{\infty} D_{2D}\left(-\frac{\partial f}{\partial E}\right) dE = q^2 \langle D_{2D}(E_F) \rangle . (2.23)$$

The factor, $-(\partial f / \partial E)$ acts like a δ-function with a width of about $k_B T_L$ at the Fermi level, so the quantum capacitance is proportional to the average density of states at the Fermi level.

Inversion layer charge below threshold:

Having obtained $Q_i(V_{GS})$ above threshold [eqn. (2.16)] we seek a corresponding relation below V_T. Below V_T $Q_i \ll Q_B$, but it is important because it carries the subthreshold current of a MOSFET. As sketched in Fig. 2.5, the subthreshold inversion layer charge is observed to vary exponentially with gate voltage. Our objective is to derive an expression that describes this behavior. Instead of (2.10) we have

$$\varepsilon_{Si} E_S = -Q_B = qN_A W_D, \qquad (2.24a)$$

where W_D is the width of the surface depletion region. Equation (2.24a) can also be expressed as

$$E_S = qN_A W_D / \varepsilon_{Si} = q N_A / C_D, \qquad (2.24b)$$

where

$$C_D \equiv \varepsilon_{Si} / W_D \qquad (2.24c)$$

is the semiconductor depletion capacitance. With eqn. (2.24b), eqn. (2.7) becomes

$$Q_i(\psi_S) = -\frac{k_B T_L}{q} C_D \left(\frac{n_i}{N_A}\right)^2 e^{q\psi_S / k_B T_L} . \qquad (2.25)$$

Equation (2.25) shows that Q_i varies exponentially with surface potential, but we seek $Q_i(V_{GS})$.

If we were to use eqn. (2.12) to relate ψ_S to V_{GS}, the result would not be pretty. Alternatively, recall that eqn. (2.16b) implies the equivalent circuit of Fig. 2.7. Voltage division gives

$$\psi_S = \frac{C_{ox}}{C_{ox}+C_D}V_G' = \frac{V_G'}{m},$$ (2.26)

where

$$m \equiv 1 + C_D/C_{ox}.$$ (2.27)

Equation (2.26) is not exactly correct, because C_D is non-linear, so the appropriate average depletion layer depth should be used. At the threshold voltage where $V_G' = V_T'$, $\psi_S = 2\psi_B$, we have

$$2\psi_B = \frac{k_B T_L}{q}\ln\left(\frac{N_A}{n_i}\right)^2 = \frac{V_T'}{m}.$$ (2.28)

Finally, using eqns. (2.26) - (2.28) in eqn. (2.25), we obtain

$$Q_i(V_{GS}) = -(m-1)C_{ox}\frac{k_B T_L}{q}e^{q(V_{GS}-V_T)/mk_B T_L}.$$ (2.29)

Equations (2.16a) and (2.29) explain the $Q_i(V_G)$ characteristic sketched in Fig. 2.5, which shows that Q_i varies exponentially with gate voltage below V_T and linearly with gate voltage above V_T. (This plot should be compared to Fig. 2.5a, which plots $Q_S(\psi_S)$.) Having established the key features of 1D MOS electrostatics, we must now consider the 2D effects that arise from the drain-to-source voltage.

V'_G

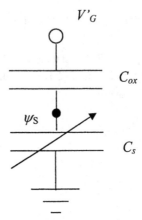

C_{ox}

ψ_S

C_S

Fig. 2.7   Illustration of voltage division for the MOS capacitor. Below threshold, $C_S =$ C_D, the depletion layer capacitance, and above threshold, $C_S = C_{inv}$, the inversion layer capacitance.

2.4   2D MOS Electrostatics

To understand a MOSFET, we need to include the effect of the drain potential and understand $Q_S(V_{GS}, V_{DS})$. Figure 2.8 is a sketch of $E_C(x,0)$ vs. x from source to drain. The equilibrium case is shown in Fig. 2.8a, which should be compared to Fig. 2.4b. The pn junctions between the source/drain and channel have a built-in potential under flatband conditions of

$$V_{bi} = \frac{k_B T_L}{q} \ln \frac{N_A N_D}{n_i^2}. \qquad (2.30)$$

As shown in Fig. 2.8a, a gate voltage can increase ψ_S and lower the barrier between the source and channel. At the threshold of inversion, $\psi_S = 2\psi_B$, and the barrier height is

$$E_b = k_B T_L \ln\left(\frac{N_D}{N_A}\right). \qquad (2.31)$$

For typical values ($N_D \cong 10^{20} cm^{-3}$ and $N_A = 10^{18}$ cm^{-3}) $E_b \approx 0.1$ eV. Above threshold, the barrier is even smaller.

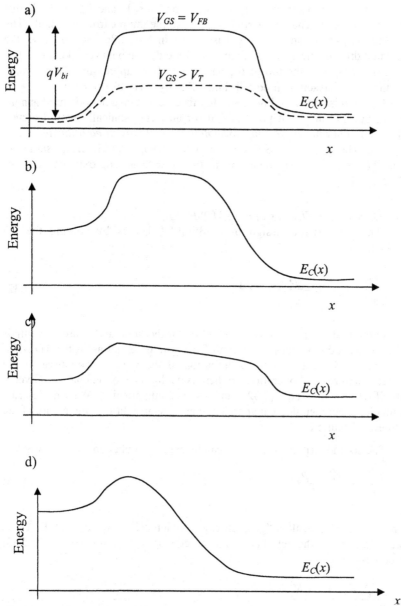

Fig. 2.8   Sketch of conduction band energy vs. position for four different bias conditions.
a)   equilibrium, $V_{GS} = V_{DS} = 0$;   b) $0 < V_{GS} < V_T, V_{DS} \gg 0$,   c)
$V_{GS} \gg V_T, V_{DS} \approx 0$, d). $V_{GS} \gg V_T, V_{DS} \gg 0$.

Figure 2.8b is a sketch of $E_C(x,0)$ vs. x for $V_{GS} < V_T$ and $V_{DS} \gg 0$. In this case, the source to channel barrier is large, so the drain current is small. The case for $V_{GS} \gg V_T$ and V_{DS} small is shown in Fig. 2.8c. In this case, the potential drop is linear in the channel, and the device behaves like a resistor. Note that charge at the top of the barrier ($x = 0$) is approximately the value for the MOS capacitor in equilibrium. Finally, $E_C(x,0)$ vs. x for $V_{GS} \gg V_T$ and $V_{DS} \gg 0$ is shown in Fig. 2.8d. In this case, the channel potential is non-linear, but the important point is that for an electrostatically well-designed MOSFET, $Q_S(x=0)$ *is still approximately what it was in equilibrium*, when $V_{DS} = 0$. The goal of MOSFET design is to manage the 2D electrostatics so that $Q_S(x = 0)$ is nearly independent of V_{DS} with an approximate value of $C_{GS}(V_{GS} - V_T)$.

Two Dimensional Electrostatics in MOSFETs

The electrostatic design of a MOSFET begins by solving the 2D Poisson equation,

$$\frac{\partial^2 \psi}{\partial x^2} + \frac{\partial^2 \psi}{\partial y^2} = -\frac{\rho}{\varepsilon_{Si}} = +\frac{qN_A}{\varepsilon_{Si}}, \tag{2.32}$$

where the first term accounts for the effect of the drain and source potentials and the second describes the effect of the gate potential as in the 1D MOS capacitor. (See Fig. 2.4 for a definition of the x and y directions.) We assume subthreshold operation, so there is negligible mobile charge. Given a MOSFET structure, eqn. (2.32) can be solved numerically. We will describe a simple, phenomenological approach that gives insight into the nature of the numerical solutions.

The second term in eqn. (2.32) can be expressed phenomenologically as

$$\frac{\partial^2 \psi}{\partial y^2} = \frac{(V'_{GS} - \psi_S)}{\Lambda^2}, \tag{2.33}$$

where Λ is the so-called "geometric scaling length" [2.2]. For small Λ, the second term on the left hand side of eqn. (2.32) dominates, and (2.33) becomes

$$\frac{(V'_{GS} - \psi_S)}{\Lambda^2} = \frac{qN_A}{\varepsilon_{Si}} \tag{2.34a}$$

or

$$V'_{GS} = \psi_S + qN_A \Lambda^2 / \varepsilon_{Si} \ . \tag{2.34b}$$

This is the one-dimensional case, where $\partial^2 \psi / \partial y^2$ dominates and $\partial^2 \psi / \partial x^2$ can be ignored. For the 1D MOS capacitor, we already found that

$$V'_{GS} = \psi_S - Q_B / C_{ox} = \psi_S + qN_A W_D / C_{ox}, \tag{2.34c}$$

so to make eqn. (2.34b) consistent with 1D MOS theory, we must have

$$\Lambda = \sqrt{\frac{\varepsilon_{Si}}{\varepsilon_{ox}} W_D t_{ox}} \ . \tag{2.35}$$

Having specified Λ, we can use eqn. (2.33) in eqn. (2.32) to find

$$\frac{\partial^2 \psi_S}{\partial x^2} - \frac{(\psi_S - V'_{GS})}{\Lambda^2} = \frac{qN_A}{\varepsilon_{Si}} \ . \tag{2.36}$$

If we define

$$\phi = \psi_S - V'_{GS} + \frac{qN_A \Lambda^2}{\varepsilon_{Si}} \ , \tag{2.37}$$

then eqn. (2.36) becomes

$$\frac{d^2 \phi}{dx^2} - \frac{\phi}{\Lambda^2} = 0, \tag{2.38}$$

which can be solved subject to the boundary conditions

$$\phi(0) = \phi_S \tag{2.39a}$$

$$\phi(L) = \phi_D \tag{2.39b}$$

to find

$$\phi(x) = Ae^{-x/\Lambda} + Be^{x/\Lambda}. \tag{2.40}$$

The final solution,

$$\psi_S(x) = (V'_{GS} - qN_A\Lambda^2/\varepsilon_{Si}) + \phi(x) \tag{2.41}$$

is sketched in Fig. 2.9. The first term on the RHS of eq. (2.41) describes the effect of the gate potential; it tries to hold ψ_S constant at a value determined by V_{GS}. The second term describes the lowering of $E_C(x)$ due to the drain and source potentials, and, if the channel length is too short compared to Λ can lead to so-called drain induced barrier lowering (DIBL), as shown in Fig. 2.9.

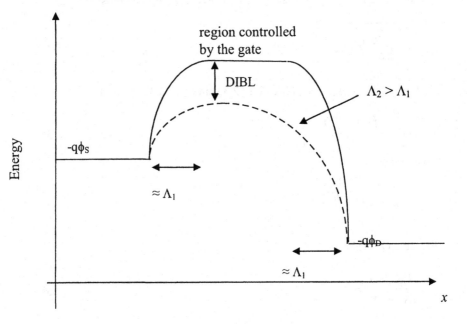

Fig. 2.9   Sketch of the electron energy vs. position showing the role of the geometric scaling length, Λ. For a given L, a short Λ (Λ_1 above) results in a region of the channel that is strongly controlled by the gate, but for a longer Λ (Λ_2 above) DIBL occurs and the gate does not have total control over the potential barrier.

To produce a portion of the channel where ψ_S is controlled by V_{GS} and is independent of V_{DS}, we require

$$L \gg \Lambda, \tag{2.42}$$

which is the criterion for an electrostatically well-designed MOSFET. An electrostatically well-designed MOSFET is one for which the channel is controlled by the gate voltage, not by the drain to source voltage. A small Λ permits the smallest L while maintaining acceptable DIBL and S.

Equation (2.35) shows that the oxide and silicon body thickness should be small for good gate control. Similar considerations apply to a single gate SOI MOSFET, with W_D replaced by the thickness of the silicon body, T_{Si}. Double gate MOSFETs have a smaller Λ [2.9] and the cylindrical gate MOSFET, an even smaller one [2.10]. A careful derivation based on a series solution to Poisson's equation gives Λ from a transcendental equation [2.10]. Figure 2.10 shows why Λ is called a geometric scaling length; it depends on the geometry of the MOSFET. In a bulk MOSFET, field lines from the drain can reach through and lower the barrier near the source. By surrounding the channel with the gate, the field lines from the drain terminate on the gate, which screens the drain potential so that it does not influence the channel potential near the source.

a)

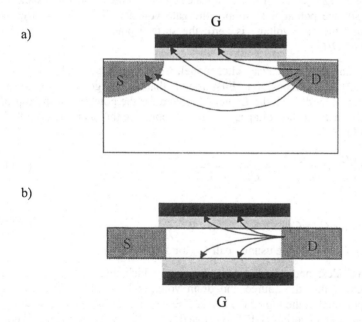

b)

Fig. 2.10   Illustration of the drain electric field lines in a) a bulk and b) a double gate MOSFET.

A Capacitor Model for 2D Electrostatics

The MOSFET's gate electrode induces charge in the channel of the transistor, but the source and drain electrodes can also induce charge. As sketched in Fig. 2.11a, our interest is in the potential and charge at the top of the barrier (which defines the beginning of the channel). The three-capacitor model of Fig. 2.11b describes the modulation of the potential and charge at the top of the barrier by the three terminals. This simple model provides an alternative, very physical, picture of 2D MOSFET electrostatics.

Before using the three capacitor model, recall the 1D results of eqn. (2.12a), which relates the charge and surface potential to the gate voltage. Solving for the surface potential, we find

$$\psi_S = V'_{GS} + \frac{Q}{C_{ox}}. \tag{2.43}$$

When the charge, Q, is zero, the Laplace solution to the electrostatics shows that the surface potential is simply the gate voltage. When the charge is present and the gate voltage is zero, the surface potential is the charging potential of Q/C_{ox}.

Returning to the three capacitor model, we can solve the problem shown in Fig. 2.11b by superposition. First assume that the charge at the top of the barrier is zero, and solve the Laplace problem for the potential at the top of the barrier, then set the voltages to zero and compute the charging potential. The result is

$$\psi_S = \left(\frac{C_G}{C_\Sigma} V_G + \frac{C_D}{C_\Sigma} V_D + \frac{C_S}{C_\Sigma} V_S \right) + \frac{Q}{C_\Sigma}, \tag{2.44}$$

where

$$C_\Sigma = C_G + C_D + C_S. \tag{2.45}$$

(Note that C_G is what we usually call C_{ox} for a MOSFET.)

A well-designed transistor is one for which the gate capacitance dominates so that the source and drain voltages have little effect on the surface potential at the top of the barrier. Equation (2.44) explains the drain-induced barrier lowering (DIBL) sketched in Fig. 2.3c and the classic short channel effects of a MOSFET as sketched in Fig. 2.3d. As the drain voltage increases, the source to channel barrier height decreases, which increases the subthreshold current (DIBL). As the channel length decreases, the drain capacitance increases, which lowers the barrier. A lower gate voltage, therefore, is needed to reduce the barrier height to the value given by eqn. (2.31), where the MOSFET turns on. The result is a decrease of V_T with decreasing channel length (threshold voltage roll-off).

(a)

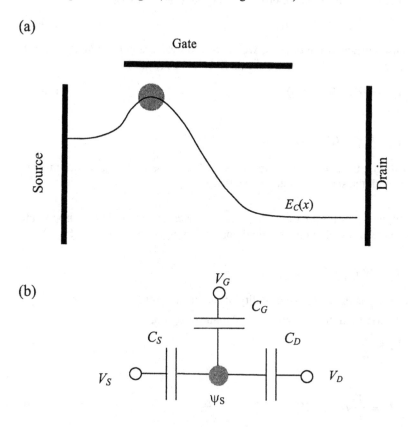

(b)

Fig. 2.11   Illustration of the capacitor model of 2D electrostatics. (a) The conduction band profile under high gate and drain bias with the inversion layer charge at the top of the barrier identified. (b) A three-capacitor model that describes the control of the potential at the top of the barrier.

2.5   MOSFET Current vs. Voltage Characteristics

Having understood the $Q_i(V_{GS}, V_{DS})$ characteristic, it is relatively easy to establish the essential features of the MOSFET $I_D(V_{GS}, V_{DS})$ characteristic. As shown in Fig. 2.12, for low V_{DS}, the MOSFET behaves like a resistor while for high V_{DS}, it behaves more like a current source. To minimize the mathematics, we will develop an expression for the linear (low V_{DS}) and saturated (high V_{DS}) regions only.

The drain current is the product of charge times velocity,

$$I_D = -W \, Q_i(x)\langle \upsilon(x)\rangle. \tag{2.46a}$$

Because current is continuous, we choose to evaluate eqn. (2.46a) at $x = 0$, where we know $Q_i(x = 0)$ from eqn. (2.16). The result is

$$I_D = -W \, Q_i(0)\langle \upsilon(0)\rangle \tag{2.46b}$$

or

$$I_D = W \, C_{GS} \left(V_{GS} - V_T\right)\langle \upsilon(0)\rangle. \tag{2.46c}$$

(Recall that the gate-source capacitance, C_{GS}, is less than C_{ox}, because it is in series with the semiconductor capacitance, C_S.)

In the linear region of the $I_D - V_{DS}$ characteristic, the potential drop in the channel is linear, the electric field approximately constant (recall Fig. 2.8c), so

$$I_D = W \, C_{ox} \left(V_{GS} - V_T\right)\mu_{eff} \mathrm{E}_x, \tag{2.47}$$

where μ_{eff} is the effective mobility of the inversion layer electrons and E_x is the electric field in the channel. In the linear region,

$$\mathrm{E}_x \approx \frac{V_{DS}}{L}, \tag{2.48}$$

so

$$I_D = \frac{W}{L} \mu_{eff} C_{ox} \left(V_{GS} - V_T\right)V_{DS} = \frac{V_{DS}}{R_{ch}}. \tag{2.49}$$

The channel resistance,

$$R_{ch} = \frac{1}{\mu_{eff} C_{ox} \left(V_{GS} - V_T\right)} \frac{L}{W}, \tag{2.50}$$

is proportional to the length of the channel, as expected.

a)   b)

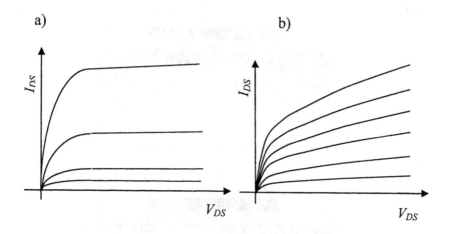

Fig. 2.12   MOSFET I_D vs. V_{DS} characteristic a) long channel (on-current varies as the
              square of the gate voltage) and b) short channel (on-current varies linearly with
              the gate voltage).

Saturation Region: Long Channel:
        When $V_{DS} > 0$, the potential along the channel varies with position. The
simplest way to treat this two-dimensional problem is to modify the one-
dimensional result, eqn. (2.16), to

$$Q_i(x) = -C_{ox}\left(V_{GS} - V_T - V(x)\right),\qquad(2.51)$$

where $V_S < V(x) < V_D$ is the channel potential. Equation (2.51) is the well
known "gradual channel approximation" of MOS theory [2.7, 2.8]. Under
low drain bias, $V(x)$ is small, and the inversion layer is uniform as sketched
in Fig. 2.13a.   When the drain bias increases, however, the potential
difference between the gate and substrate is reduced near the drain, so the
inversion layer density decreases. When

$$V(x = L) = V_{Dsat} = \left(V_{GS} - V_T\right),\qquad(2.52)$$

then eqn. (2.51) predicts that $Q_i(x = L) = 0$. As shown in Fig. 2.13b, the
channel is said to be "pinched-off" at $x = L$. (There is, of course, a small,
positive Q_i to carry the drain current. The value is determined by the velocity
of carriers in the pinched-off region.)

(a)

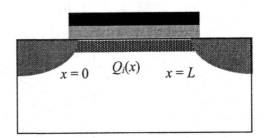

(b)

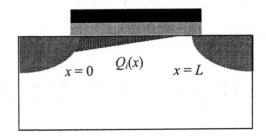

Fig. 2.13   Illustration of the channel for $V_{GS} > V_T$ and (a) low and (b) high V_{DS} conditions.

When $V_{DS} > V_{Dsat}$, the potential drop across the inverted portion of the channel is $\approx (V_{GS} - V_T)$, so we can estimate the electric field as

$$E_x(0) \approx \frac{(V_{GS} - V_T)}{2L} \tag{2.53}$$

(the factor of 2 comes from a proper treatment of the non-uniform electric field within the channel). Finally, from eqns. (2.47) and (2.53), we find the saturation current as

$$I_{Dsat} = \frac{W}{2L} \mu_{eff} C_{ox} (V_{GS} - V_T)^2. \tag{2.54}$$

The square law behavior of the long channel MOSFET was sketched in Fig. 2.12a.

Saturation Region: Short Channel

For long channel MOSFETs we can assume that $\langle v \rangle = \mu E_x$, but at high electric fields, transport becomes nonlinear as was sketched in Fig. 1.11. When the field is higher than about 10^4 V/cm, the velocity of electrons in silicon saturates at about 10^7 cm/s. Such fields occur for nanoscale MOSFETs with $L \approx 100$nm and $V_{DS} \approx 1.0$V. For a short channel MOSFET, therefore, eqn. (2.46c) gives the saturated drain current as

$$I_D = W C_{ox} v_{sat} (V_{GS} - V_T). \tag{2.55}$$

Figure 2.12 compared a long channel MOSFET for which $I_D \propto (V_{GS} - V_T)^2$ to a short channel, velocity saturated MOSFET for which $I_D \propto (V_{GS} - V_T)$. In practice, $I_D \propto (V_{GS} - V_T)^\alpha$, where $1 < \alpha < 2$.

Finally, we should mention that transport across a short, high-field region (the channel of a nanoscale MOSFET under high bias) is one of the more difficult problems in transport theory. Figure 1.12 sketched the steady state $\langle v \rangle$ vs. x profile for a high field step. The velocity saturates at $\approx 10^7$ cm/s, but initially it overshoots. The near-equilibrium carriers injected into the high-field region initially have high mobility. As they gain energy from the field, they scatter more, and the mobility decreases. Velocity overshoot occurs because the mobility and electric field are both high near the beginning of the step. The spatial extent of velocity overshoot is roughly 100nm, so for a nanoscale MOSFET, velocity overshoot may occur throughout the entire channel.

Subthreshold Conduction:

Under subthreshold conditions, the source to channel barrier is large, Q_i is small, and the electric field in the channel is also small. If diffusion dominates,

$$\langle v(0) \rangle = D_{eff} / L, \tag{2.56}$$

so, using eqn. (2.29) for $Q_i(V_{GS})$, eqn. (2.46b) gives

$$I_D = \frac{W}{L}(m-1)\mu_{eff}C_{ox}\left(\frac{k_B T_L}{q}\right)^2 e^{q(V_{GS}-V_T)/mk_B T_L}. \tag{2.57}$$

Figure 2.14 sketches the $I_D(V_{GS})$ characteristic below and above V_T. In the subthreshold region, I_D varies exponentially with $(V_{GS} - V_T)$, but above threshold, it varies as $(V_{GS}-V_T)^\alpha$. The subthreshold swing is readily evaluated from eqn. (2.57) to give

$$S = 2.3\,m\,\frac{k_B T_L}{q} \qquad \text{(V/decade)} \qquad\qquad (2.58)$$

Since $m > 1$ (recall that $m = 1 + C_D/C_{ox}$), we find that S > 60 mV/decade. Finally, note that

$$I_{off} = I_D(V_{GS}=0, V_{DS}=V_{DD}) \propto e^{-qV_T/mk_B T_L} \qquad\qquad (2.59a)$$

and

$$I_{on} = I_D(V_{GS}=V_{DS}=V_{DD}) \propto (V_{DD}-V_T)^\alpha \qquad\qquad (2.59b)$$

so

$$I_{off} \propto e^{\beta I_{on}^{1/\alpha}}, \qquad\qquad (2.59c)$$

where β is a constant. The result is that if we lower V_T to get a little more on-current, we get exponentially more off-current. In fact, technologies are often characterized by a $\log_{10} I_{off}$ vs. I_{on} plot.

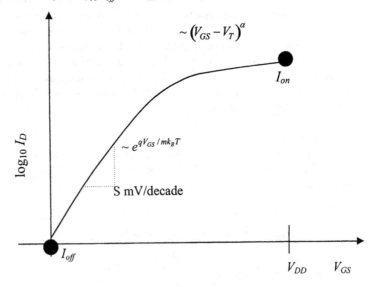

Fig. 2.14   The $\log_{10} I_D$ vs. V_{GS} characteristic illustrating the important parameters below and above threshold.

2.6   The Bipolar Transistor

Figure 2.15 compares the idealized structures and energy band diagrams for MOS and bipolar transistors. For the bipolar transistor, the base-emitter voltage lowers the emitter-base energy barrier, so that $n(0)$ electrons are injected into the base. The injected carriers diffuse across the base and are collected by the collector. Since the density of electrons is low at the base-collector junction, we find

$$J_C = qD_n \frac{dn}{dx} = qn(0)\frac{D_n}{W_B} = \left[q\,\frac{n_i^2}{N_A}\frac{D_n}{W_B} \right]e^{qV_{BE}/k_BT_L}, \qquad (2.60)$$

where the second expression comes from the well-known "Law of the Junction" for $n(0)$ [2.8]. The collector current follows directly to write

$$I_C = qA_E \frac{n_i^2}{N_A}\langle\upsilon(0)\rangle e^{qV_{BE}/k_BT_L}, \qquad (2.61)$$

where $\langle\upsilon(0)\rangle$ is the diffusion velocity and A_E the emitter area. Equation (2.61) describes the bipolar transistor in the normal, active mode of operation where the emitter-base junction is forward-biased and the base-collector junction reverse biased.

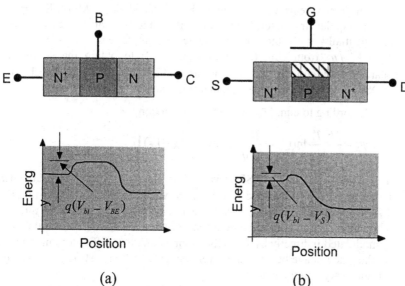

(a)   (b)

Fig. 2.15   Idealized device structures and energy band diagrams for: (a) bipolar transistor and (b) a MOSFET.

Because of the similarity of the bipolar and MOS transistor energy band diagrams as displayed in Fig. 2.15, we should expect that they operate similarly. Let's see if we can derive the MOSFET characteristic from eqn. (2.61), which describes the bipolar device.

By recognizing that ψ_S plays the role of V_{BE}, eqn. (2.61) becomes

$$I_C = qWt_{inv}\frac{n_i^2}{N_A}\langle v(0)\rangle e^{q\psi_S/k_BT_L}, \tag{2.62}$$

where Wt_{inv} is the cross-sectional area for current flow. Recall that

$$n(0) = \left(\frac{n_i^2}{N_A}\right)e^{q\psi_S/k_BT_L} \tag{2.63}$$

and that $n_S = t_{inv}n(0)$, so

$$I_D = WQ_i(0)\langle v(0)\rangle, \tag{2.64}$$

which is precisely eqn. (2.46b) from which we derived the MOSFET $I_D(V_{GS}, V_{DS})$ characteristic.

One often hears the statement "below threshold, a MOSFET operates like a bipolar transistor," by which it is meant that the current varies exponentially with the input voltage. This is easy to see by using $\psi_S = V'_{GS}/m$ from eqn. (2.26) in eqn. (2.62) to obtain the subthreshold characteristic, eqn. (2.57). It is not as well known that above threshold, the MOSFET still operates like a bipolar transistor [2.11].

According to eqn. (2.11), in strong inversion,

$$\psi_S = \frac{2k_BT_L}{q}\ln(Q_i) \approx \frac{2k_BT_L}{q}\ln[C_{ox}(V_{GS}-V_T)]. \tag{2.65}$$

The drain current of a MOSFET varies exponentially with ψ_S both above **and** below threshold, but above threshold, ψ_S varies logarithmically with $(V_{GS} - V_T)$ so the result is that I_D varies linearly with $(V_{GS} - V_T)$ above threshold. The reduced control of the gate occurs because the source to channel barrier is modulated indirectly by V_G. Above threshold, it is difficult to modulate ψ_S by the gate voltage because the inversion layer charge is strong and it screens out the charge on the gate.

2.7 CMOS Technology

The important device performance metrics are derived from the requirements of CMOS circuits. The basic element of a CMOS digital system is the inverter shown in Fig. 2.16. It consists of two normally off (so-called enhancement mode) MOSFETs in series. The one on the top is a p-channel MOSFET ($V_T < 0$) and the one on the bottom is an n-channel MOSFET ($V_T > 0$). The PMOS transistor is referred to as the "pull-up" transistor, because when it is on, it pulls the output up to the power supply voltage. Similarly, the NMOS transistor is referred to as the "pull down" transistor, because when it is on, it pulls the output voltage down to ground. If MOSFETs were ideal switches that closed when the gate voltage exceeded V_T (or is less than V_T for the PMOS) then the transfer characteristic (the output voltage vs. input voltage) would be the dashed line in Fig. 2.16b. When the input voltage is low, the PMOS transistor turns on (NMOS off) and connects the output node to V_{DD}; when the input voltage is high, the NMOS transistor turns on (PMOS off) and connects the output node to ground. If a low voltage represents a logical 0 and a high voltage a logical 1, then this circuit operates as a digital inverter. More complicated logical functions can be realized by placing transistors in series and parallel.

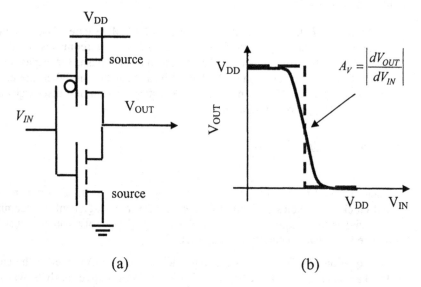

(a)   (b)

Fig. 2.16   A CMOS inverter. (a) circuit schematic and (b) the transfer characteristic. (The dashed line is the transfer characteristic if the CMOS transistors were ideal switches.)

The solid line in Fig. 2.16b is the transfer characteristic of a well-designed inverter. The switching point occurs at $V_{DD}/2$, which is achieved by matching the currents and threshold voltages of the two transistors. (PMOS transistors have lower mobility than NMOS transistors, so the PMOS transistor is typically 2-3 times wider than the NMOS transistor.) The slope of the transfer characteristic at the switching point is the voltage gain, A_V. (By biasing a transistor at this point, the inverter can be used as an analog amplifier.) Note that away from the switching point, the output voltage is insensitive to the input voltage, which provides the inverter with a *noise margin*. Even if a logical zero is not exactly 0 volts (because of noise on the input line), the output voltage will be restored to a logical 1. Similarly, a logical 1 may be less than V_{DD} volts, and the output voltage is restored exactly to a logical 0. Noise margins are what make digital systems possible, otherwise noise would degrade the signals after propagating through a few stages. In order to have noise margins at the low and high end (that is regions of the transfer characteristic that are flat at low and high input voltages), we require

$$|A_V| > 1. \tag{2.66}$$

Gain provides signal restoration, which is essential for any device to be used in a digital system.

Figure 2.17 shows the pull down portion of a CMOS gate. The capacitor, C, represents the capacitance of the gates and interconnects that are connected to the output node. During the pull-up phase, the capacitor is charged to V_{DD} – a logic one. A clock drives the gate, and when the clock voltage is high, the NMOS pull-down transistor turns on and discharges the capacitor. The power dissipation is

$$P = \frac{1/2\,CV_{DD}^2}{T_{CL}/2} = fCV_{DD}^2, \tag{2.67}$$

where T_{CL} is the period of the clock. As the number of transistors per chip and clock frequencies continue to increase, power management is becoming a crucial issue. Equation (2.67) explains why the supply voltage must be reduced with each technology generation.

Equation (2.67) describes the dynamic (or switching) power of the gate. The key advantage of CMOS, which led to its adoption over NMOS, was the fact that essentially no power was dissipated unless a gate switched. But this is changing as leakage currents are increasing. Consider the circuit of Fig.

2.17 when the NMOS pull down is off and C is charged. The static power dissipated is

$$P_S = I_{off}V_{DD}. \tag{2.68}$$

To minimize the static power, a low I_{off} (high V_T) is required, but a high V_T reduces the on-current and the speed suffers as discussed next.

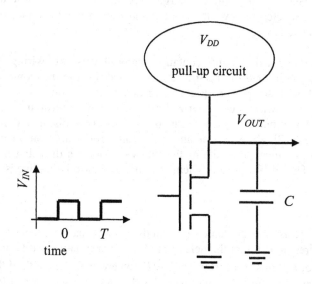

Fig. 2.17 Illustration of the pull-down portion of a CMOS gate.

The gate switching delay determines the maximum clock frequency. If we measure the gate delay by the time, τ, to remove the capacitor's charge, we find

$$\tau = \frac{CV_{DD}}{I_{on}}, \tag{2.69}$$

which explains why a high on-current is important. From eqn. (2.46c), with we can evaluate the device delay metric as

$$\tau = \frac{C_{GS} W L V_{DD}}{W C_{GS}(V_{GS} - V_T)\langle\upsilon(0)\rangle} \approx \frac{L}{\langle\upsilon(0)\rangle} \tag{2.70}$$

so the device delay metric is closely related, but not identical to, the transit time of carriers across the channel. The delay metric for current-day technology (the 90nm technology node) is about 1 ps. This represents the intrinsic switching speed of a transistor and would correspond to a clock frequency of several hundred GHz. Integrated circuits operate much slower because of the need to charge and discharge large capacitances because the output node is connected to several gates (fan-out) and the interconnecting wires also add capacitance.

A typical integrated circuit contains many layers of wiring to interconnect the circuit, and the need to charge and discharge interconnects limits speed and increases power. Figure 2.18 is a cross sectional sketch of an interconnect. The wire is characterized by a capacitance per unit length and a resistance per unit length and can be viewed as a distributed RC transmission line [2.7]. The mathematics of signal propagation on an RC transmission line is identical to the mathematics of particles diffusing in a semiconductor. The "diffusion coefficient" for the RC transmission line is

$$D_{RC} = 1/R_w C_w , \tag{2.71a}$$

where R_w is the resistance per unit length and C_w the capacitance per unit length of the wire. Recall that the delay time for minority carriers diffusing across the base of a bipolar transistor is $W^2/2D$, where W is the width of the base. By analogy, therefore, the time for a signal to propagate across an RC transmission line is

$$\tau_{RC} = 0.5 R_w C_w L_w^2, \tag{2.71b}$$

where L_w is the length of the interconnect. The resistance per unit length depends on the resistivity of the metal interconnect, which explains why copper metallization has replaced aluminum metallization. The capacitance per unit length depends on the geometry of the interconnect and the dielectric constant of the medium, which explains why so-called low-κ insulators are replacing SiO_2 in the interconnect layers. The critical point, however, is that the delay depends on the *square* of the length of the interconnect. (Note, however, that for short interconnects, eqn. (2.71b) could imply that signals propagate faster than the speed of light). For such cases, a more physical transmission line model must be used.

An integrated circuit contains a distribution of interconnect lengths, most of which are short, local interconnects. A few long interconnects, however, are essential, and these long interconnects limit the speed of the system. Originally, device speed determined the speed of the circuit, but now interconnect delays dominate, and a central task of the chip designer is to manage these delays.

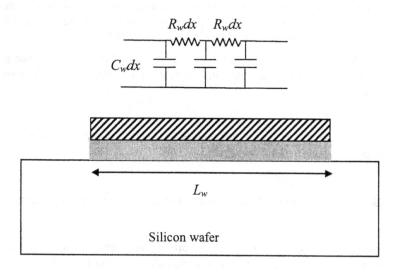

Fig. 2.18   Cross-sectional sketch of an interconnect wire showing how it is modeled as a distributed RC transmission line.

MOSFET Scaling:

The objective of device scaling is to shrink transistor dimensions so that more transistors can be placed on a chip. Typically, the scaling factor, k, is $\approx \sqrt{2}$, so that the area of a transistor shrinks by one-half and the number per area increases by a factor of two. The challenge in scaling is to maintain suitable electrical characteristics.

Consider scaling the channel length,

$$L \rightarrow L/k. \tag{2.72}$$

To control two-dimensional effects such as DIBL and V_T roll-off, we require that $L >> \Lambda$, where Λ is the geometric scaling length discussed in Sec. 2.4. In practice

$$L > 1.5\Lambda \tag{2.73}$$

provides acceptable control of two-dimensional electrostatics. It is necessary, therefore, that we also scale Λ,

$$\Lambda \rightarrow \Lambda/k, \tag{2.74}$$

which, according to eqns. (2.35) can be accomplished by reducing t_{ox} and W_D (by increasing the channel doping). The source/drain junction depth, an effect not included in eqns. (2.35) should also be reduced.

Because scaling increases the number of transistors per chip, the power dissipation per chip increases unacceptably unless the power supply voltage is also scaled,

$$V_{DD} \rightarrow V_{DD}/k. \tag{2.75}$$

The on-current per unit width, however, must be maintained so that circuit speed does not suffer. Therefore,

$$I_{ON}/W \rightarrow I_{ON}/W. \tag{2.76}$$

Because V_{DD} is reduced, we must also reduce V_T to maintain on-current,

$$V_T \rightarrow V_T/k, \tag{2.77}$$

but eqn. (2.59a) shows that the off-current, and therefore the standby power, increases exponentially as V_T is scaled down. The on-current/off-current trade-off is an increasingly difficult challenge to manage. Note also that as V_T is scaled down, variations in V_T increase for small devices, so device-to-device variations are becoming an important issue.

It is also interesting to see how channel resistance of the MOSFET scales. We find,

$$R_{ch} \equiv \frac{V_{DD}}{I_{ON}} \rightarrow R_{ch}/k. \tag{2.78}$$

The intrinsic resistance of the device is scaling down, but the parasitic resistances, which depend on metal-semiconductor contact resistance and the junction depth and doping, are increasing. The result is that device performance will be increasingly degraded by series resistance as channel lengths push into the nanoscale regime.

Moore's Law states that the number of transistors per chip doubles each technology generation [2.12]. (Currently, a technology generation is about two years). The doubling of transistor density is a result of three factors: 1) device scaling, 2) improvements in layout, which increase transistor packing density, and 3) increased die (chip) size. Because the scaling factor, k, is somewhat greater than $1/\sqrt{2}$, factors 2) and 3) play an important role in Moore's Law. The International Technology Roadmap for Semiconductors is a statement of the technology characteristics needed to maintain Moore's Law in the future [2.13]. Current technologies have channel lengths below 100nm, and if scaling continues for another decade or so, channel lengths will be less than 10nm. Designers are increasingly challenged by off-state leakage (from the source to drain and through the ultra-thin gate insulator), low on-currents, increasing variation of device parameters across a chip, interconnect delays, and power dissipation.

2.8   Ultimate Limits

We conclude this chapter with a look at the ultimate limits for transistor-based digital computation. In particular, we seek to establish the:

1) Minimum energy dissipation per logic transition, E_{min} (J)

2) Minimum device size, L, in nm (or maximum device density, n_S, per cm^2)

3) Minimum device delay, t_{min} (ps)

4) Power dissipation, P (W/cm^2).

The topic of dissipation in computing is a deep problem [2.14], [2.15] with a rich history [2.16]. Some of these issues are still being debated. Our goal in this section is modest; we seek to establish the ultimate limits for any transistor that operates in a conventional circuit by modulating the flow of current across an electrostatic barrier. We follow an approach that is similar to [2.17] and [2.18].

To begin, let's consider the switching energy, the energy dissipated when we convert a logical one to a logical zero,

$$E_S = \frac{1}{2}CV_{DD}^2 = \frac{1}{2}QV_{DD} = \frac{q}{2}NV_{DD}, \qquad (2.79)$$

where Q is the charge stored on the capacitor, and N is the number of electrons stored. The minimum switching energy occurs when $N = 1$, so we need to establish the minimum power supply voltage, V_{min}.

According to eqn. (2.66), a CMOS inverter must have a gain greater than one in order to have a noise margin. We compute the gain by equating I_D(NMOS) to I_D(PMOS), solving for V_{OUT}, and differentiating it with respect to V_{IN}. If we assume subthreshold conduction, ideal devices ($m = 1$ in eqn. (2.57)), and symmetrical N and P-channel devices, then we find that [?]

$$V_{min} = 2\ln(2)k_B T_L / q.$$
(2.80)

When eqn. (2.80) is inserted into eqn. (2.79), we find that the minimum energy per logical transition is [2.17]

$$E_S = \ln(2)k_B T_L \quad (T_L = 300K).$$
(2.81)

Equation (2.81) is a specific example of a much more general result. Whenever a logical transition occurs and the bit is erased, then there is an inevitable energy dissipation that must be at least as great as the result of eqn. (2.81) [2.14]. If the capacitors are charged and discharged adiabatically, however, or if information is preserved (so-called reversible computing), then the energy dissipation per logic transition can be arbitrarily small [2.15]. The interested reader is referred to the literature for these topics. Our focus is on the limits of conventional CMOS logic.

Having addressed question 1), we now turn to questions 2) – 4) and use an approach that is similar to that of Zhirnov, *et al.* [2.18]). Consider the "transistor" sketched in Fig. 2.19, which consists of two thermal equilibrium reservoirs (source and drain) separated by an energy barrier whose height is modulated by a gate voltage. We assert that the results of this analysis apply in general to any transistor that controls a current by modulating a potential energy barrier. Since our focus is on limits, we assume a ballistic transistor. No scattering occurs in the channel – only in the source/drain regions where strong scattering maintains thermal equilibrium.

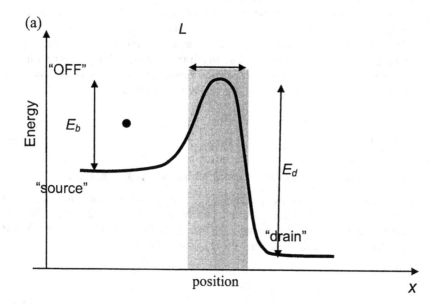

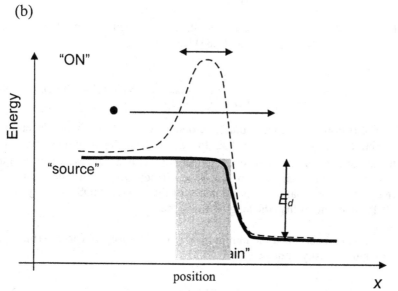

Fig. 2.19   Illustration of a hypothetical, digital electronic device. (a) the off-state and (b) the on-state.

Consider the switching energy for the model transistor. In the off-state, an electron injected from the source must have less than a 50:50 chance of propagating to the drain, and in the on-state, an electron injected into the drain dissipates its energy in the drain and thermalizes. There must be a minimum barrier (drain voltage) so that a thermal equilibrium electron in the drain has less than a 50:50 chance of returning to the source (this assures us that we can distinguish the two states). We conclude that

$$e^{-E_d/k_B T_L} < \frac{1}{2},$$

(2.82)

which leads to the same minimum switching energy as eqn. (2.81).

Next, we address the question of the minimum size of a transistor. In order for the off-state to be distinguishable from the on-state, we require that the tunneling probability through the barrier be less than 0.5. Using a WKB approximation for the tunneling probability, we find

$$P(\text{WKB}) = \exp\left(-\frac{2\sqrt{2mE}}{h} L \right) < \frac{1}{2}.$$

(2.83)

If we assume a thermal equilibrium electron in the source, $E = k_B T = E_{min}/\ln(2)$, then eqn. (2.83) leads to

$$L > \frac{[\ln(2)]^{3/2}}{2}\left(\frac{\hbar}{\sqrt{2mE_S}} \right) \approx \left(\frac{\hbar}{\sqrt{2mE_S}} \right) = 1.5 \text{ nm } (T_L = 300\text{K}).$$

(2.84)

The last form is what we could have obtained directly from the Uncertainty Principle, $\Delta x \, \Delta p \geq \hbar$. According to eqn. (2.84), the minimum size of a transistor is about 1 nm at room temperature. MOSFETs with gate lengths only a few times larger than this have already been demonstrated [2.19]. (Note, however, that the minimum size of the overall MOSFET is typically 10-15 times larger than the length of the gate.)

Having determined the minimum size of a device, we can determine the maximum density of devices as

$$n_{max} = \frac{1}{L_{min}^2} \approx 4.7 \times 10^{13} \text{ devices / cm}^2,$$

(2.85)

which is an enormous number – four to five orders of magnitude larger than present day CMOS. Two orders of magnitude come from the fact that a device is at least 10 times larger than its minimum feature. But more importantly, as we shall see later, the density of devices is not limited by our ability to make small devices; it is limited by our ability to dissipate the power generated by the devices.

Consider next the switching speed of the transistor, which is the transit time across the control region,

$$t_S = \frac{L}{\upsilon} = \frac{L}{\sqrt{2E/m}}. \tag{2.86}$$

Using $L = L_{min}$ and $E = k_B T_L = E_{min}/\ln(2)$, we find

$$t_{min} = \left(\frac{\ln(2)}{2}\right)^{5/2} \frac{\hbar}{E_{min}} \approx \frac{\hbar}{E_{min}} = 40 \text{ fs}, \tag{2.87}$$

which is only a few times smaller than CMOS transistors are expected to achieve.

Finally, let us estimate the power dissipation per cm² for a chip operating at the density and speed limits. We have

$$P = \frac{\alpha n_S E_S}{t_S}, \tag{2.88}$$

where α is the switching activity factor (the average fraction of clock cycles that an average transistor switches). If we assume that $\alpha = 1$ and use the minimum switching energy and maximum switching speed we find

$$P_{max} = \frac{n_{max} E_{min}}{t_{min}} = 3.7 \times 10^6 \text{ W/cm}^2. \tag{2.89}$$

Equation (2.89) is almost three orders of magnitude higher than the energy flux at the surface of the sun! Silicon technologists hope to be able to develop affordable heat sinking techniques to dissipate 100 W/cm², but that is more than four orders of magnitude smaller than the density-limited power dissipation given by eqn. (2.89). Power dissipation is already a critical issue for designers. Present day technology operates at four to five orders of magnitude above $k_B T \ln(2)$. Operating speeds are well below fundamental limits, but the switching energy is much larger than $k_B T \ln(2)$. Operation at relatively a relatively high voltage (~ 1V) is necessary to minimize errors due to spontaneous thermal transitions, to accommodate device-to-device variations, and to provide sufficient speed. Our ability to remove thermal energy from the chip, not our ability to make devices small, is what limits the device density of a chip. It is likely that CMOS technology will be capable of placing more transistors on a chip than can be tolerated from a power dissipation point of view.

We have established rather optimistic upper limits for transistors. For example, we assumed an on-off ratio of 2. Realistic designs require much higher ratios. We also assumed an operation voltage of $~k_BT$, but such a low voltage would result in too many spontaneous errors. New transistor materials and structures make allow CMOS technology to operate closer to these fundamental limits, but no barrier-modulation transistor can be fundamentally better than the silicon MOSFET.

2.9   Summary

This chapter summarized the conventional theory of MOS devices and circuits. It provides some background for examining the MOSFET from a new perspective. We seek a new understanding of small MOSFETs, one that explains the key results for submicron MOSFETs that we have just reviewed, but that also applies all the way to the scaling limit. We also seek an approach that applies to MOSFETs, as well as to the unconventional devices that are being explored to complement or even replace the MOSFET. One important point that will arise again is the central importance of self-consistent electrostatics. The device performance metrics and circuits and systems aspects of speed and power will also apply to any device intended for use in the conventional digital systems that we are familiar with.

Chapter 2 References

[2.1]   J. D. Meindl, Q. Chen, and J. A. Davis, "Limits on Silicon Nanoelectronics for Terascale Integration," *Science*, **293**, pp. 2044-2049, 2001.

[2.2]   D. Frank, et al., "Device Scaling Limits of Si MOSFETs and Their Application Dependencies," *Proc. IEEE*, **89**, pp. 259-288, 2001.

[2.3]   M. Ieong, B. Doris, J. Kedzierski, K. Rim, and M. Yang, "Silicon Scaling to the Sub- 10-nm Regime," *Science*, **306**, pp. 2057-2060, 2004.

[2.4]   X. Huang, et al., "Sub-50 nm P-channel FinFET," *IEEE Trans. Electron. Dev.*, **48**, pp. 880-886, 2001.

[2.5]   B.S. Doyle, S. Datta, M. Doczy, S. Hareland, B. Jin, J. Kavalieros, T. Linton, A. Murthy, R. Rios, and R. Chau, "High Performance Fully-Depleted Tri-Gate CMOS Transistors," *IEEE Electron Dev. Lett.*, **24**, pp. 263-265, 2003.

[2.6]   B. Gobel, et al., "Fully Depleted Surrounding Gate Transistor (SGT) for 70nm DRAM and Beyond," *Tech. Digest, International Electron Devices Meeting,* San Francisco, CA, Dec. 9-11, 2002.

[2.7]   Y. Taur and T. Ning, *Fundamentals of Modern VLSI Devices*, Cambridge Univ. Press, Cambridge, U.K., 1998.

[2.8]   R.F. Pierret, *Fundamentals of Semiconductor Devices*, Addison-Wesley, Reading, MA, 1996.

[2.9]   K. Suzuki, T. Tanaka, Y. Tosaka, H. Horie, and Y. Arimoto, "Scaling Theory of Double-Gate SOI MOSFETs," *IEEE Trans. Electron Dev.*, **40**, pp. 2326—2329, 1993.

[2.10] C.P. Auth and J.D. Plummer, "Scaling Theory for Cylindrical, Fully-Depleted, Surrounding-Gate MOSFET's," *IEEE Electron Dev. Lett.*, **18**, pp. 74-77, 1997.

[2.11] E. O. Johnson, "The Insulated-Gate Field-Effect Transistor - A Bipolar Transistor in Disguise," *RCA Review*, **34**, pp. 80-94, 1973.

[2.12] G. E. Moore, "Cramming more components onto integrated circuits," Electronics, 38, pp. 114-117, 1965

[2.13] http://public.itrs.net

[2.14] R. Landauer, "Irreversibility and Heat Generation in the Computing Process," *IBM J. Res. and Dev.*, **5**, pp. 183-191 (1961).

[2.15] C.H. Bennet, "Logical Reversability of Computation," *IBM J. Res. and Dev.*, **17**, pp. 525-532 (1973).

[2.16] S. Datta, Quantum Transport Atom to Transistor, 1st Ed., Cambridge University Press, Cambridge, UK. 2005.

[2.17]   J. D. Meindl and J.A. Davis, "The Fundamental Limit on Binary Switching Energy for Terascale Integration (TSI)," *IEEE J. Solid-State Circuits,* **35**, pp. 1515-1516, 2000.

[2.18]   V. V. Zhirnov, R. K. Cavin, J. A. Hutchby, G. I. Bourianoff, "Limits to Binary Logic Switch Scaling – A Gedanken Model," *Proc. of the IEEE,* Special Issue on Nanoelectronics and Nanoscale Processing, Bing Sheu, Peter Wu & Simon Sze, Guest Editors, Nov., 2003.

[2.19]   B. Doris, M. Ieong, T. Kanarsky, Y. Zhang, R.A. Roy, O. Dokumaci, Z. Ren, F-F Jamin, L. Shi, W. Natzle, H-J Huang, J. Mezzapelle, A. Mocuta, S. Womack, M. Gribelyuk, E. C. Jones, R. J. Miller, H-S P. Wong, and W. Haensch, "Extreme Scaling with Ultra-Thin Si Channel MOSFETs," Tech. Dig., IEEE Electron Devices Mtg., pp. 267-270, Washington, Dec. 2002.

Chapter 3: The Ballistic Nanotransistor

3.1 Introduction

Silicon MOSFETs with channel lengths less than 10nm long have now been realized [3.1], and at such dimensions, our traditional understanding of transistors has to be questioned. In this chapter, we describe an approach to MOSFETs based on concepts that are widely-used in mesoscopic physics [3.2]. Figure 3.1 shows the lowest conduction subband energy vs. position along the channel of a 10nm MOSFET under a variety of gate and drain biases [3.3]. The figure shows that there is an energy barrier between the source and channel and that the gate voltage modulates the height of the barrier. The drain current increases as the barrier height is reduced by the increasing gate voltage (Fig. 3.1a for low drain bias and Fig. 3.1b for high drain bias). For an electrostatically well designed MOSFET, the gate bias controls the height of the source to channel barrier; the drain bias has a small effect (Figs. 3.1c and 3.1d). Because current flow is controlled by injection across a barrier, a MOSFET is similar to a bipolar transistor, except that in a bipolar transistor the barrier height is directly modulated by the emitter-base voltage while in the MOSFET it is modulated indirectly by the gate voltage [3.4]. Our theory of the nanoscale MOSFET will be based on this simple, physical picture.

Before proceeding, we should clarify what is being plotted in Fig. 3.1. Electrical engineers plot "energy band diagrams," the conduction band minimum (or valence band maximum) vs. position. In nanoscale devices, however, quantum confinement effects are strong, so it is more appropriate to plot the electron eigenstates. Figure 3.2 shows the difference for a double gate MOSFET of the type mentioned in Chapter 2. Because quantum confinement in the z-direction is strong, the subbands are widely separated and do not couple. We can, therefore, solve a one-dimensional Schrödinger equation in the z-direction at various positions along the channel in the x-direction. Figure 3.2 shows the result. Note that the conduction band minima, $E_C(x, y)$, varies in two dimensions, but the subband profiles vary only in the x-direction. Unless otherwise noted, all energy band diagrams will refer to plots of the subband energy vs. position – not the actual conduction band minima.

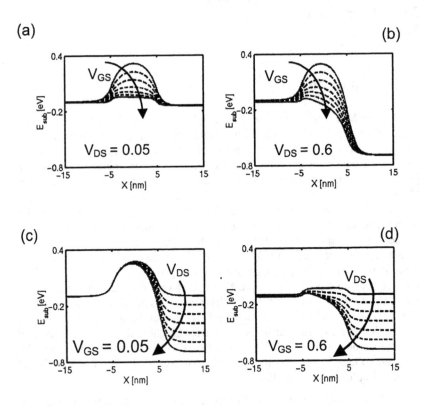

Figure 3.1   Energy band diagrams (conduction subband vs. position) for a 10nm channel length, double gate, n-channel MOSFET under a variety of bias conditions. (Reproduced with permission from [3.3])

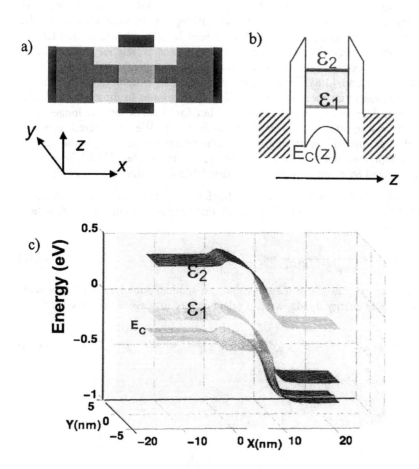

Figure 3.2   Comparison of the conduction band profile and the subband profile. (a) the
double gate device structure. (b) illustration of the quantum confinement in
the z-direction, (c) the conduction band and subband profiles in the x-y plane.
(Courtesy of Ramesh Venugopal.)

3.2 Physical View of nano-MOSFETs

MOSFETs are complicated by 2D electrostatics and strong, off-equilibrium transport in the presence of rapidly varying electric fields. These effects have been recently examined by 2D, numerical simulations using the non-equilibrium Green's function approach (see Sec. 1.9 and Refs. [3.5, 3.6]; see also [3.7] for an earlier quantum scale study of 10 nm MOSFETs and [3.8] for a recent review). For extremely short channel lengths, quantum mechanical tunneling from source to drain (through the barrier) degrades device performance, but for channel lengths longer than about 10 nm, MOSFETs behave classically [3.3]. We will, therefore, adopt a classical model. Scattering, which complicates the analysis, is the subject of Chapter 4, but the essential physics of nanoscale MOSFETs can be established by examining numerical simulations of ballistic MOSFETs.

Let's first examine the carrier distribution function within a nanoscale MOSFET. Before doing that, recall that in equilibrium, the distribution function is

$$f_0(E) = \frac{1}{1 + e^{(E-E_F)/k_B T_L}} \approx e^{(E_F - E)/k_B T_L}, \qquad (3.1a)$$

where the second form assumes that the semiconductor is nondegenerate. For simple energy bands, we can relate energy to wavevector or velocity to write

$$f_0(k) \approx e^{(E_F - E_C)/k_B T_L} \times e^{-\hbar^2 k^2 / 2m^* k_B T_L} = e^{(E_F - E_C)/k_B T_L} \times e^{-m^* v^2 / 2k_B T_L}. \qquad (3.1b)$$

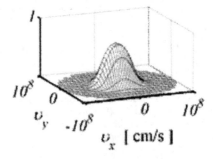

Figure 3.3. Equilibrium electron velocity distribution function for two-dimensional, nondegenerate electrons. Under degenerate conditions this shape changes from the isotropic Maxwellian shown here to an isotropic Fermi-Dirac distribution. (Reproduced with permission from [3.9])

For two-dimensional carriers like those in the inversion layer of a MOSFET, the equilibrium carrier distribution function is a symmetrical Maxwellian (or Fermi-Dirac) distribution as shown in Fig. 3.3. The question is: What does the distribution function look like within the channel of a ballistic MOSFET?

As discussed in Sec. 1.7, it is straightforward to solve the Boltzmann equation for the carrier distribution function within a ballistic device. Figure 3.4 shows the numerically computed carrier distribution function vs. position for a 10nm channel length ballistic MOSFET operating under high bias [3.9]. The carrier distribution function is seen to be strongly distorted from its equilibrium value. Note, for example, the ballistic peak that develops within the channel.

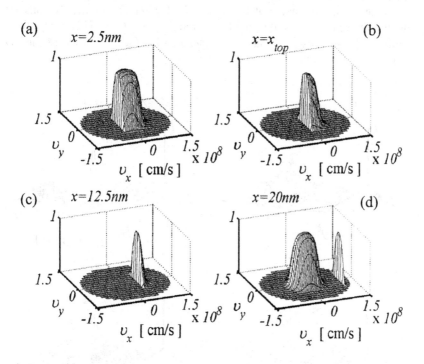

Figure 3.4. Electron distribution function vs. position for a ballistic n-MOSFET as computed under high gate and drain bias [3.9]. (a) in the n+ source, (b) at the top of the barrier, (c) deep within the channel, and (d) within the n+ drain. For the n+ source, 0<x<7.5nm, for the intrinsic channel, 7.5<x<17.5nm, and for the n+ drain, 17.5<x<25nm. (Reproduced with permission from [3.9])

The complex shape of the carrier distribution function would seem to preclude simple, analytical modeling, but Fig. 3.5 shows that things are much simpler at the top of the source-channel barrier because at this point, all positive k-states are occupied according to the source Fermi level and all negative k-states according to the drain Fermi level. Figure 3.5 shows the carrier distribution function at the top of the barrier under high gate bias at four different drain voltages. For $V_{DS} = 0$, the distribution has an equilibrium shape; the positive half of the thermal distribution was injected from the thermal reservoir at the source and the negative half from the drain. As V_{DS} increases, the negative portion of the distribution diminishes. Note, however, that the positive half <u>grows</u>. This occurs because of the MOSFET's electrostatics. In an electrostatically well-designed MOSFET, the gate holds the charge at the top of the barrier approximately constant with drain bias. Above threshold, the value is $\approx C_{ox}(V_{GS} - V_T)$, except for a shift of the threshold voltage, V_T, depending on the magnitude of the 2D short channel effects.

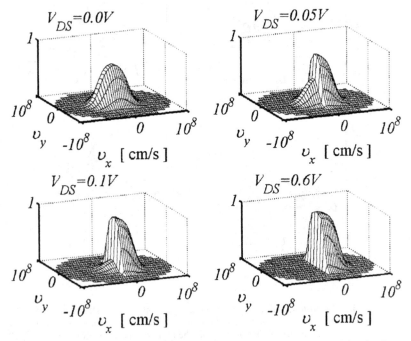

Figure 3.5.   Electron distribution function at the top of the source-channel barrier for a ballistic n-MOSFET as computed under high gate and four different drain biases. (Reproduced with permission from [3.9])

For a ballistic MOSFET, the average electron velocity within the channel varies rapidly in space and reaches very high values. Due to the lack of scattering in the channel, at high drain biases, the carrier velocity near the drain end of the channel can be much higher than the carrier saturation velocity

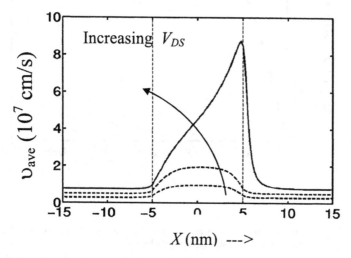

Figure 3.6.   Average electron velocity vs. position for the ballistic, double gate n-MOSFET under high gate bias and at three different drain biases. (Reproduced with permission from [3.3])

We have already seen in Fig. 3.5 that the electron density at the top of the barrier is approximately independent of drain bias, so it is natural to inquire how the average velocity at the top of the barrier varies with drain bias. As shown in Fig. 3.7, the average velocity initially increases with V_{DS}, then saturates. The magnitude of the saturated velocity increases with gate voltage. Velocity saturation (recall Fig. 1.11) is a familiar concept in short channel MOSFETs where it occurs in the high field region where carriers are excited to high kinetic energies and scatter frequently [3.10]. Its occurrence in a ballistic MOSFET where there is no scattering, and the location of the position at which the velocity saturates (the top of the barrier where the electric field is zero) is, at first, a surprise. The explanation follows directly from Fig. 3.5. As the drain voltage increases, the drain Fermi level decreases, which lowers the population of the negative velocity k-states at the top of the barrier and increases the net velocity of the ensemble. When the negative side of the distribution is completely suppressed, the net velocity saturates at the average velocity of an equilibrium, hemi-Fermi-Dirac distribution.

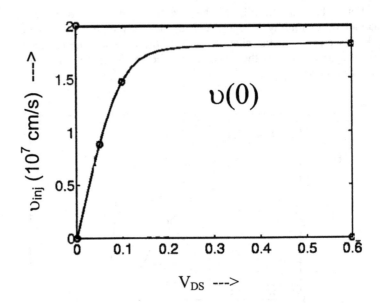

Figure 3.7.   Average electron velocity at the top of the source-channel barrier as a function of drain bias with a high applied gate bias [3.9].

The key concepts illustrated by these semi-classical numerical simulations are that:

1) the carrier distribution function at the top of the source-channel barrier consists of two thermal equilibrium halves, one injected from the source and the other from the drain,

2) as the drain bias increases, the negative portion of the distribution diminishes in size, but for an electrostatically well-designed MOSFET, the total carrier density at the top of the barrier is maintained at an approximately constant value, and

3) since the high V_{DS} distribution approaches a hemi-Fermi-Dirac distribution, the average velocity at the top of the barrier saturates at a limiting value which is the average velocity of a thermal equilibrium hemi-Fermi-Dirac distribution.

As discussed next, these concepts can be used to develop an analytical theory of the ballistic MOSFET.

3.3 Natori's Theory of the Ballistic MOSFET

In this section, we present an overview of the theory of the ballistic MOSFET without the mathematical details that will be discussed in the following section. Figure 3.8 sketches the E-k relation at the top of the source-channel barrier. The positive k-states are populated by injection from the source according to the source Fermi level, E_F, and the negative k-states are populated from the drain according to the drain Fermi level, $E_F - qV_{DS}$. For $V_{DS} = 0$, the positive and negative states are equally occupied, the average velocity is zero, and $I_D = 0$. For small V_{DS}, the drain Fermi level is lowered, fewer negative k-states are occupied, the net velocity is positive, and $I_{DS} \propto V_{DS}$. For large V_{DS}, the negative k-states are empty, the average velocity saturates at a value determined by the gate voltage, which determines the location of the Fermi level within the conduction band. In this case the drain current saturates at the so-called on-current.

We can relate the carrier populations in the positive and negative halves to their respective Fermi levels. Above threshold, the electrostatics of a well-designed MOSFET then demand that the total carrier density is approximately independent of drain voltage, so we find

$$n_S^+(E_F) + n_S^-(E_F - qV_{DS}) = Q(0)/(-q) \approx C_{ox}(V_{GS} - V_T)/(-q), \tag{3.2}$$

where n_S^+ and n_S^- are given by eqns. (1.43a) and (1.43b). Equation (3.2) is an equation for the location of the Fermi level. For a given device design, the gate capacitance, C_{ox}, and threshold voltage, V_T, are determined. Equation (3.2) then determines the location of the Fermi level as a function of gate and drain bias. As the drain bias increases, n_S^- decreases, so E_F must increase to maintain charge balance. Physically, this occurs by the gate electrostatically pushing down the source-channel barrier to let more electrons in from the source, and it explains why the positive half of the distribution in Fig. 3.5 grows as V_{DS} increases.

Having determined the Fermi level, the positive and negative fluxes can be evaluated by integrating over the populated states (i.e. by computing directed moments) to obtain the drain current as

$$I_D = W\left[J^+(E_F) - J^-(E_F - qV_{DS})\right], \tag{3.3a}$$

which can also be written as

$$I_D = qW\left[n_S^+(E_F)\upsilon^+(E_F) - n_S^-(E_F - qV_{DS})\upsilon^-(E_F - qV_{DS})\right]. \tag{3.3b}$$

a)

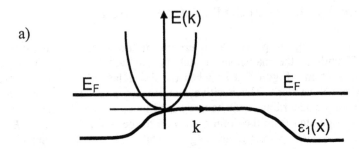

b)

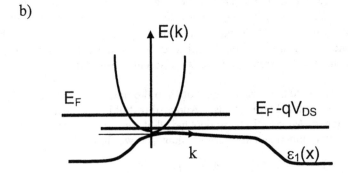

c)

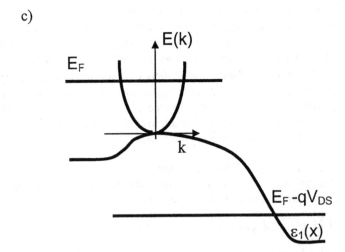

Figure 3.8   *E-k* relation at the top of the source-channel barrier showing the source and
              drain Fermi levels, E_F and $E_F - qV_{DS}$. A high gate voltage, which sets the
              location of the Fermi level in the conduction subband, is assumed. a) $V_{DS} = 0$,
              b) V_{DS} small, and c) V_{DS} large.

Equations (3.2) and (3.3b) can be solved to find

$$I_D = W Q(0) \upsilon^+ \frac{\left[1 - \left(n_S^-/n_S^+\right)\left(\upsilon^-/\upsilon^+\right)\right]}{\left[1 + n_S^-/n_S^+\right]}. \tag{3.4}$$

The gate voltage controls $Q(0)$ through MOS electrostatics, and the source and drain voltages control the carrier densities and velocities. In the following section, we will evaluate Eqn. (3.4) to obtain the I_{DS} vs. V_{DS} characteristics of a ballistic MOSFET.

The I_D vs. V_{DS} characteristics for a ballistic MOSFET at room temperature are sketched in Fig. 3.9. Note that they are quite similar to those of a scattering dominated MOSFET. Our objective is to evaluate the $I_D(V_{GS}, V_{DS})$ characteristic, but we shall also focus on understanding the low V_{DS} channel conductance, the high V_{GS}, V_{DS} on-current, and the drain saturation voltage.

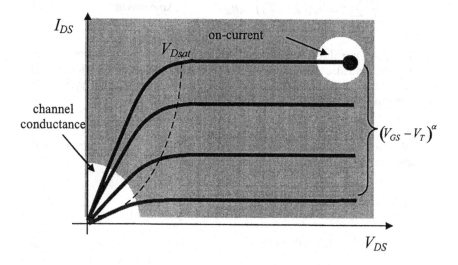

Figure 3.9   Sketch of the I_{DS} vs. V_{DS} characteristic of a ballistic MOSFET.

3.4 Nondegenerate, Degenerate, and General Carrier Statistics

To evaluate Natori's theory of the ballistic MOSFET, we need to evaluate the parameters in Eqn. (3.4). As discussed in Sec. 1.7, evaluating these directed moments is particularly easy at the top of the barrier, where we need to evaluate them. Specifically, we must evaluate

$$n_S^+(0) = \frac{1}{A} \sum_{k_y, k_x > 0} f_0(E_F) \qquad\qquad \text{cm}^{-2} \qquad\qquad (3.5a)$$

$$n_S^-(0) = \frac{1}{A} \sum_{k_y, k_x < 0} f_0(E_F - qV_D) \qquad\qquad \text{cm}^{-2} \qquad\qquad (3.5b)$$

$$J^+ = \frac{1}{A} \sum_{k_y, k_x > 0} q \upsilon_x f_0(E_F) \equiv q n_S^+ \upsilon^+ \qquad\qquad \text{Ampere/cm} \qquad\qquad (3.5c)$$

$$J^- = \frac{1}{A} \sum_{k_y, k_x < 0} q \upsilon_x f_0(E_F - qV_D) \equiv q n_S^- \upsilon^- \qquad\qquad \text{Ampere/cm} \qquad\qquad (3.5d)$$

In the following three subsections, we will evaluate these expressions to obtain the ballistic I-V characteristic for three different conditions: i) nondegenerate carrier statistics, ii) fully degenerate carrier statistics (i.e. T_L= 0K) and iii) general conditions.

3.4.1 The Ballistic MOSFET (nondegenerate conditions)

We first derive the *I-V* characteristic assuming that the electron gas is nondegenerate, In this case, the sums over k-space work out to be

$$n_S^+(0) = \left(\frac{N_{2D}}{2}\right) e^{(E_F - \varepsilon_1(0))/k_B T_L} \qquad\qquad (3.6a)$$

$$n_S^-(0) = \left(\frac{N_{2D}}{2}\right) e^{(E_F - qV_D - \varepsilon_1(0))/k_B T_L} \qquad\qquad (3.6b)$$

$$\upsilon^+ = \upsilon^- = \upsilon_T = \sqrt{\frac{2k_B T_L}{\pi m^*}} \qquad\qquad (3.6c)$$

$$I^+ = W J^+ = W q n_S^+ \upsilon_T \qquad\qquad (3.6d)$$

$$I^- = W J^- = W q n_S^- \upsilon_T, \qquad\qquad (3.6e)$$

where N_{2D} is the two-dimensional effective density-of-states as given by Eqn. (1.23), and we have assumed that a single subband, at energy, $\varepsilon_1(0)$, is occupied. The appropriate effective mass to use depends on the subband. Assuming that only the lowest subband is occupied, the appropriate effective mass for (100) silicon is $m^* = m_t = 0.19\,m_0$, which gives a thermal velocity of $\upsilon_T = 1.2 \times 10^7$ cm/s at room temperature, T_L=300K.

From eqns. (3.6a) and (3.6b), we see that $n_S^-/n_S^+ = e^{-qV_D/k_B T_L}$, so Eqn. (3.4) becomes

$$I_D = q\,W\,\upsilon_T\,n_S(0)\frac{\left(1-e^{-qV_D/k_B T_L}\right)}{\left(1+e^{-qV_D/k_B T_L}\right)},\qquad(3.7a)$$

which, using MOS electrostatics for $q\,n_S = C_{ox}(V_G - V_T)$, gives the I-V characteristic of the ballistic MOSFET under nondegenerate conditions as

$$I_D = W\,C_{ox}\upsilon_T\,(V_G - V_T)\frac{\left(1-e^{-qV_D/k_B T_L}\right)}{\left(1+e^{-qV_D/k_B T_L}\right)}.\qquad(3.7b)$$

Although the assumption of nondegenerate carrier statistics is a poor one above threshold, Eqn. (3.7b) is a simple expression that provides some insight into ballistic nanotransistors. A plot of the common source characteristics from Eqn. (3.7b) gives characteristics like those sketched in Fig. 3.9. The drain current of a ballistic MOSFET (under nondegenerate conditions) saturates when $V_D > V_{Dsat} \approx 2(k_B T_L/q)$. For large drain voltages, we find the on-current as

$$I_D(on) = W\,C_{ox}\upsilon_T\,(V_G - V_T),\qquad(3.8)$$

which has the form of a traditional, velocity-saturated MOSFET model [3.15], except that υ_{sat} is replaced by the thermal injection velocity, υ_T. The characteristic exponent for the gate voltage dependence, is $\alpha = 1$, just as it is in the traditional velocity saturation model. These similarities to the traditional velocity saturation model should have been expected because we saw that the average carrier velocity does saturate at the charge control point located at the top of the barrier.

Figure 3.9 and Eqn. (3.7b) also show that the ballistic MOSFET has a finite channel conductance. For small drain voltages, the exponentials can be expanded to find

$$I_D = \left[W\, C_{ox}\left(V_G - V_T\right) \frac{\upsilon_T}{2\left(k_B T_L / q\right)} \right] V_D = G_{CH} V_D, \qquad (3.9)$$

where G_{CH} is the channel conductance. We have assumed that only one subband is populated, but the number of available modes in the transverse direction is proportional to the width, W, of the MOSFET. Alternatively, we can express the channel conductance as

$$G_{CH} = I_D(\text{on})/(2 k_B T_L / q). \qquad (3.10)$$

It is also instructive to compare the ballistic channel conductance to the traditional, drift-diffusion result from Eqn. (2.49),

$$G_{CH} = W\, C_{ox}\left(V_G - V_T\right) \frac{\mu_{eff}}{L_{eff}}. \qquad (3.11)$$

The conductance of a MOSFET cannot be greater than the ballistic limit, so we conclude that the traditional model is valid as long as

$$\mu_{eff} \frac{\left(2 k_B T_L / q\right)}{L_{eff}} \ll \upsilon_T. \qquad (3.12)$$

Finally, we note that the origin of the ballistic conductance (analogous to Landauer's quantum conductance) is easy to appreciate in this nondegenerate picture. Consider Fig. 3.10, which sketches the energy band diagram under high gate bias and low drain bias. At the top of the barrier, we have a positive current, I^+, injected from the source. Because those carriers injected from the drain experience a larger barrier, I^- is smaller by a factor of $e^{-q V_D / k_B T_L}$, so the net current is

$$I_D = I^+\left(1 - e^{-q V_D / k_B T_L}\right) \approx I^+\left[V_D / (k_B T_L / q)\right], \qquad (3.13)$$

where the second form follows for small V_D. Under low drain bias, $I^+ = W\left[Q(0)/2\right]\upsilon_T$ whereas under high drain bias, $I^+ = W\, Q(0)\upsilon_T$, so I^+ under low drain bias is $I_D(\text{on})/2$ and Eqn. (3.13) reduces to Eqn. (3.10).

The ballistic conductance is a direct consequence of our thermionic emission model as evaluated under a low drain voltage assumption.

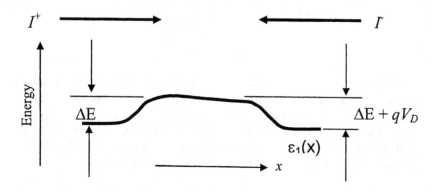

Figure 3.10   The energy band diagram of a MOSFET under high gate voltage and low drain bias showing the ballistic currents injected from the source, I^+, and from the drain, I^-.

3.4.2   The Ballistic MOSFET ($T_L = 0$, degenerate conditions)

At $T_L = 0$K, the directed moments in Eqns. (3.5) can be readily evaluated to find:

$$n_S^+(0) = \frac{k_F^2}{4\pi} = \frac{m^*}{2\pi\hbar^2}\left[E_F - \varepsilon_1(0)\right]\Theta(E_F - \varepsilon) \tag{3.14a}$$

$$n_S^-(0) = \frac{k_{FD}^2}{4\pi} = \frac{m^*}{2\pi\hbar^2}\left[E_F - qV_D - \varepsilon_1(0)\right]\Theta(E_F - qV_D - \varepsilon_1(0)) \tag{3.14b}$$

$$\upsilon^+ = \frac{4\upsilon_F}{3\pi} = \frac{4}{3\pi}\sqrt{\frac{2\left[E_F - \varepsilon(0)\right]\Theta(E_F - \varepsilon_1(0))}{m^*}} \tag{3.14c}$$

$$\upsilon^- = \frac{4\upsilon_{FD}}{3\pi} = \frac{4}{3\pi}\sqrt{\frac{2\left[E_F - qV_D - \varepsilon(0)\right]\Theta(E_F - qV_D - \varepsilon_1(0))}{m^*}} \tag{3.14d}$$

$$I^{+} = \frac{qW\hbar k_F^3}{3\pi^2 m^*} = qW \frac{\left[2m^*\left[E_F - \varepsilon(0)\right]\Theta\left(E_F - \varepsilon_1(0)\right)\right]^{3/2}}{3m^*\pi^2\hbar^2} \quad (3.14e)$$

$$I^{-} = \frac{qW\hbar k_{FD}^3}{3\pi^2 m^*} = qW \frac{\left[2m^*\left[E_F - qV_D - \varepsilon(0)\right]\Theta\left(E_F - qV_D - \varepsilon_1(0)\right)\right]^{3/2}}{3m^*\pi^2\hbar^2} \quad .(3.14f)$$

Note that at $T_L = 0K$, +k states are occupied only when $E_F > \varepsilon(0)$ and –k states only when $E_F - qV_D > \varepsilon(0)$, which is why we have introduced the step functions, Θ. Using Eqns. (3.14) in Eqn. (3.4), we obtain the *I-V* characteristic for the ballistic MOSFET at $T_L = 0K$. In this subsection, we will discuss only three quantities: 1) the channel conductance, G_{CH}, 2) the on-current, I_D(on), and 3) the drain saturation voltage, V_{Dsat}.

To evaluate the channel conductance, we note that the directed fluxes depend on the Fermi level and assume that $V_D << V_{Dsat}$. In this region, I^+ and I^- are nearly equal so

$$I^{-} = I^{+} - \left(\frac{\partial I^{+}}{\partial E_F}\right)qV_D, \quad (3.15)$$

from which, we obtain

$$I_D = \left(I^{+} - I^{-}\right) = \left(\frac{\partial I^{+}}{\partial E_F}\right)qV_D. \quad (3.16)$$

Using Eqn. (3.14e), we evaluate the derivative and find

$$I_D = \left[\left(\frac{W\,k_F}{\pi}\right)\frac{2q^2}{h}\right]V_D. \quad (3.17)$$

Our derivation assumes that the transistors width, *W*, in the y-direction is large so that we can integrate over the transverse modes. As illustrated in Fig. 3.11, we impose periodic boundary conditions so that the allowed k-states in the y-direction are spaced by $2\pi/W$. For a given maximum wave vector in the y-direction, k_F, the number of transverse k-states that are occupied is

$$M = \frac{2k_F}{2\pi/W} = \frac{W\,k_F}{\pi} \quad (3.18)$$

so the channel conductance of a ballistic MOSFET at $T_L = 0$, from Eqn. (3.17), assumes a familiar form

$$G_{CH} = M \frac{2q^2}{h}. \tag{3.19}$$

(Note that $2k_F$ is the range of transverse k-states that are occupied, and $2\pi/W$ is the spacing of the k-states. We do not multiply by two for spin because we have already done so in Eqn. (3.17).)

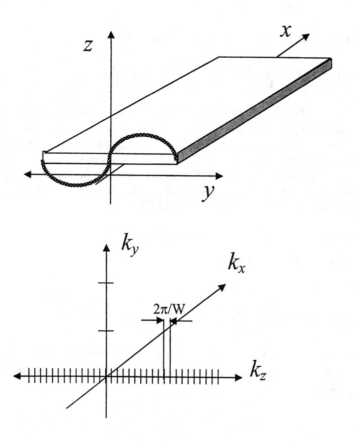

Figure 3.11   Illustration of how period boundary conditions are applied in the transverse (z) direction (i.e. the MOSFET width) and the resulting quantization of k-states in the z-direction

The transistor's on current consists of I^+ alone, so from Eqn. (3.14e)

$$I_D(on) = q\,W\,\frac{\hbar k_F^3}{3\pi^2 m^*},\qquad(3.20)$$

Recall that for a two-dimensional carrier gas, $n_s^+ = k_F^2/4\pi$ (Sec. 1.5) and that only positive k-states are occupied in the on-state, so

$$n_s = n_s^+ = \frac{k_F^2}{4\pi} = \frac{C_{ox}(V_G - V_T)}{q}\qquad(3.21)$$

Using Eqn. (3.21) in (3.20), we find the on-current as

$$I_D(on) = W\,C_{ox}\left[(8\hbar/3m^*)\sqrt{C_{ox}(V_G - V_T)/q\pi}\right](V_G - V_T),\qquad(3.22)$$

which shows that the characteristic exponent that describes the gate voltage dependence of the on-current is $\alpha = 3/2$. In the previous section, we saw that $\alpha = 1$ for nondegenerate conditions, so ballistic MOSFETs will display a characteristic exponent of $1 \le \alpha \le 3/2$. This is the same range of exponents observed for sub-micron MOSFETs that presumably operate far below the ballistic limit, so the gate voltage dependence doesn't provide a clear signature for identifying a ballistic MOSFET.

It is also useful to write the on-current as

$$I_D(on) = W\,(qn_s(0))\langle \upsilon(0)\rangle\qquad(3.23)$$

which can be re-expressed as

$$I_D(on) = W\,C_{ox}\left(\frac{4\upsilon_F}{3\pi}\right)(V_G - V_T).\qquad(3.24)$$

From the corresponding expression for the nondegenerate case, Eqn. (3.8), we see that for degenerate conditions, the gate-voltage-independent υ_T is replaced by,

$$\langle \upsilon(0)\rangle = \frac{4\upsilon_F}{3\pi} = \frac{8\hbar}{3m^*}\sqrt{\frac{C_{ox}(V_{GS} - V_T)}{q\pi}},\qquad(3.26)$$

which is both greater than υ_T and also gate-voltage-dependent,

The 3/2 power characteristic of the gate voltage dependence of the on-current arises because the carrier density varies as $(V_{GS} - V_T)$ and the carrier velocity as the square root of the same quantity.

We can also relate the $T_L = 0$ on-current to the channel conductance by using Eqn. (3.17) and (3.24) to write

$$G_{CH} = \frac{I_D(on)}{\left[2\sqrt{2}\left(E_F - \varepsilon_1(0)\right)/3q \right]}. \tag{3.27}$$

From the corresponding expression for the nondegenerate case, Eqn. (3.10), we see that for degenerate conditions,

$$\frac{2k_B T_L}{q} \rightarrow \frac{2\sqrt{2}\left[E_F - \varepsilon_1(0)\right]}{3q}, \tag{3.28}$$

where $\left[E_F - \varepsilon(0)\right]$ is evaluated under on-state conditions.

At $T_L = 0K$, the drain current saturates when the drain Fermi level drops below the conduction band. Figure 3.12 shows the location of the Fermi level under equilibrium conditions ($V_D = 0$) and for $V_D = V_{Dsat}$ For $V_D = 0$, both positive and negative k-states are occupied and the location of the Fermi level is determined from

$$n_S(0) = \frac{C_{ox}\left(V_G - V_T\right)}{q} = D_{2D} \times \left[E_F - \varepsilon_1(0)\right], \tag{3.29}$$

where D_{2D} is the two dimensional density of states. For drain voltages beyond V_{Dsat}, only positive k-states are occupied, so the location of the Fermi level is determined from

$$n_S(0) = \frac{C_{ox}\left(V_G - V_T\right)}{q} = \frac{D_{2D}}{2} \times \left[E_F - \varepsilon_1'(0)\right]. \tag{3.30}$$

The result is that $\left[E_F - \varepsilon_1'(0)\right]$, the location of the source Fermi level under high drain bias, is approximately twice its value in equilibrium. (We are assuming an electrostatically well designed MOSFET in which the inversion layer density at the top of the barrier is approximately constant with drain bias.) The mechanism for pushing the Fermi level higher into the subband for higher drain voltages is an electrostatic one. The source to channel barrier lowers to allow more carriers in from the source.

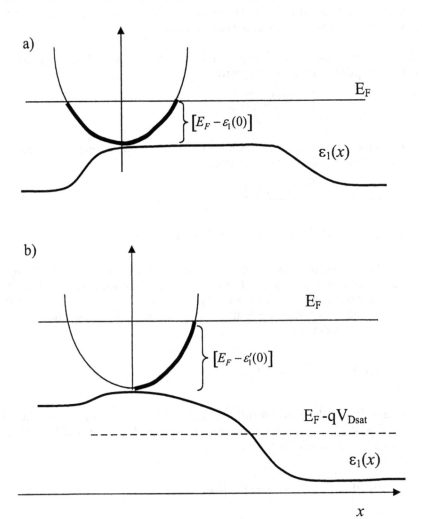

Fig. 3.12   Illustration of how the Fermi level location, $\left[E_F - \varepsilon_1(0)\right]$, and the source-channel barrier height differ at (a) $V_D = 0$ (a) and (b) $V_D = V_{Dsat}$.

Using these ideas, we see that drain current saturation occurs when $qV_D = [E_F - \varepsilon_1'(0)]$, so

$$V_{Dsat} = \left(\frac{C_{ox}}{q^2(D_{2D}/2)}\right)(V_G - V_T) = \left(\frac{2C_{ox}}{C_S}\right)(V_G - V_T),$$ (3.31)

where the semiconductor, or quantum capacitance, $C_S = q^2 D_{2D}$, was defined in Eqn. (2.21).

For a long channel MOSFET, the drain saturation voltage is $(V_G - V_T)$. The prefactor in Eqn. (3.31) is typically much less than one, so for a ballistic MOSFET at $T_L = 0K$, the drain saturation voltage is much smaller than for a long channel MOSFET. Finally, since the drain saturation voltage is simply $[E_F - \varepsilon_1'(0)]/q$, Eqn. (3.31) can be used to show

$$V_{Dsat} = 1.5 \times \left(\frac{I_D(on)}{G_{CH}}\right),$$ (3.32)

which gives some indication of how the I-V characteristic rounds off as saturation is approached.

3.4.3 The Ballistic MOSFET (General Conditions)

Having established the important features of ballistic MOSFETs, we are ready to derive a general expression for arbitrary levels of carrier degeneracy. We begin again with Eqns. (3.5) and find

$$n_s^+(0) = \frac{N_{2D}}{2} \log\left(1 + e^{(E_F - \varepsilon_1)/k_B T_L}\right) = \frac{N_{2D}}{2} F_0(\eta_F)$$ (3.33a)

$$n_s^-(0) = \frac{N_{2D}}{2} \log\left(1 + e^{(E_F - qV_D - \varepsilon_1)/k_B T_L}\right) = \frac{N_{2D}}{2} F_0(\eta_F - U_D)$$ (3.33b)

$$\upsilon^+ = \upsilon_T \frac{F_{1/2}(\eta_F)}{F_0(\eta_F)}$$ (3.33c)

$$\upsilon^- = \upsilon_T \frac{F_{1/2}(\eta_F - U_D)}{F_0(\eta_F - U_D)}$$ (3.33d)

$$I^+ = q W n_s^+ \upsilon^+.$$ (3.33e)

$$I^- = q W n_s^- \upsilon^-,$$ (3.33f)

In these expressions, F_0 is the Fermi-Dirac integral of order 0 [3.15] $F_{1/2}$ of order 1/2, $\eta_F = (E_F - \varepsilon_1)/k_B T_L$, and $U_D = V_D /(k_B T_L /q)$. Using these expressions, we find

$$\frac{n_S^-(0)}{n_S^+(0)} = \frac{F_0(\eta_F - U_D)}{F_0(\eta_F)} .$$
(3.34a)

and

$$\frac{v^-}{v^+} = \frac{F_{1/2}(\eta_F - U_D)}{F_{1/2}(\eta_F)} \frac{F_0(\eta_F)}{F_0(\eta_F - U_D)} .$$
(3.34b)

Using these results in Eqn. (3.4), we find the general I-V characteristics as

$$I_D = qWn_S \vartheta_T \left[\frac{1 - F_{1/2}(\eta_F - U_D)/F_{1/2}(\eta_F)}{1 + F_0(\eta_F - U_D)/F_0(\eta_F)} \right],$$
(3.35)

where

$$\vartheta_T = \sqrt{\frac{2k_B T_L}{\pi m^*}} \frac{F_{1/2}(\eta_F)}{F_0(\eta_F)}$$
(3.36)

is the so-called injection velocity. The location of the Fermi level, η_F, is determined by how much charge the gate induces in the semiconductor

$$C_{ox}(V_{GS} - V_T) = \frac{qN_{2D}}{2} [F_0(\eta_F) + F_0(\eta_F - U_D)].$$
(3.37)

Equations (3.35) – (3.37) specify the $I_D(V_{GS}, V_{DS})$ characteristics., but our simple model has ignored two-dimensional electrostatics; their treatment is discussed in the next section. Under nondegenerate conditions, Fermi-Dirac integrals of all orders reduce to exponentials, e^η, so Eqn. (3.35) reduces to Eqn. (3.7). For strong carrier degeneracy, the results of Sec. 3.4.2 are approached.

A sharp drain saturation voltage does not exist, but it is useful to examine the results for the on-current and channel conductance and to compare them to previous results. For high drain bias, the factor in square brackets reduces to unity, and we find the on-current as

$$I_D(on) = W C_{ox} \vartheta_T (V_G - V_T),$$
(3.38)

which is identical to the nondegenerate expression, Eqn. (3.8), except for the implicit gate bias dependence in ϑ_T. Similarly, by expanding Eqn. (3.35) for small drain voltages, we find the channel conductance and can write it as

$$G_{CH} = \left[W\,C_{ox}\,(V_G - V_T)\frac{\upsilon_T}{2(k_B T_L/q)} \right]\left(\frac{F_{-1/2}(\eta_F)}{F_0(\eta_F)} \right), \tag{3.39}$$

which is identical to Eqn. (3.9) except for the correction factor for carrier degeneracy. Finally, we have seen that there is a close relation between the ballistic on-current and the ballistic channel resistance. In the general case, Eqn. (3.35) can be expanded for small V_{DS}, and with Eqn. (3.38) we find

$$G_{CH} = \frac{I_D(\text{on})}{(2k_B T_L/q)}\left(\frac{F_{-1/2}(\eta'_F)F_0(\eta'_F)}{F_{1/2}(\eta'_F)F_0(\eta_F)} \right), \tag{3.40}$$

which is identical to Eqn. (3.10) except for the appearance of a degeneracy correction factor. In Eqn. (3.40), the primed quantities are evaluated under high drain bias and the unprimed ones under low drain bias.

3.5 Beyond the Natori Model

The simple model we have discussed illustrates several important points about ballistic MOSFETs, but there are two important simplifications that were made. First, we neglected two-dimensional electrostatics, which are important in short channel MOSFETs, and second, we *assumed* that above threshold, the charge at the top of the barrier is approximately independent of drain bias. Let's examine the second assumption first.

3.5.1 Role of the Quantum Capacitance
Assuming 1D electrostatics, the gate voltage is the sum of the surface potential and the voltage drop across the oxide (recall Eqn. (2.12a),

$$V'_G = \psi_S - Q/C_{ox}, \tag{3.41}$$

where $V'_G = V_G - V_{FB}$. We can write Eqn. (3.41) as

$$V'_G = \frac{-\varepsilon_i(0)}{q} + \frac{qn_S}{C_{ox}} \tag{3.42}$$

or as

$$\varepsilon_1(0) = -qV_G' + \frac{q^2 n_S}{C_{ox}}. \tag{3.43a}$$

Equation (3.43a) relates the energy at the top of the barrier to the gate voltage and to the charge at the top of the barrier. A positive gate voltage pushes the energy band down and reduces the source to channel barrier, but if charge is present at the top of the barrier, it wants to float up. The resulting potential at the top of the barrier is a balance between these two effects.

The charge at the top of the barrier is also related to the potential at the top of the barrier according to Eqns. (3.33a) and (3.33b) as

$$n_S = \frac{N_{2D}}{2} \left\{ F_0 \big[(E_F - \varepsilon_1(0))/k_B T_L \big] + F_0 \big[(E_F - qV_D - \varepsilon_1(0))/k_B T_L \big] \right\}. \tag{3.43b}$$

For a given gate and drain bias, Eqns. (3.43a) and (3.43b) define a nonlinear equation for $n_S(V_G, V_D)$ Conventional MOS theory makes a simplifying assumption to solve these equations.

Consider the solution to Eqn. (3.43) for two cases. First, assume that C_{ox} is small, then according to Eqn. (3.43a), a small change in n_S has a large effect on $\varepsilon_1(0)$. Now consider what happens if we increase the drain voltage. According to Eqn. (3.43b), this will reduce n_S, but according to Eqn. (3.43a), a smaller n_S will reduce $\varepsilon_1(0)$, then according to Eqn. (3.43b), the smaller $\varepsilon_1(0)$ will, in turn, increase n_S. There is, therefore, a feedback mechanism that tends to keep n_S fixed with drain bias. This was the basis for our assumption in the Natori model that n_S was set by the gate voltage and independent of the drain voltage.

Consider now the second case when C_{ox} is large. In this case, Eqn. (3.43a) shows that $\varepsilon_1(0)$ will be set directly by the gate voltage and that the charging energy has little effect. For a fixed $\varepsilon_1(0)$, Eqn. (3.43b) shows that N decreases by a factor of two as the drain voltage increases from zero to a large value.

These arguments show that for small C_{ox}, the gate voltage controls the charge at the top of the barrier (which is independent of the drain voltage) and for large C_{ox}, the gate voltage controls the potential at the top of the barrier, which is independent of the drain voltage. We refer to the first case as the "MOS limit" and the second case as the "bipolar limit." The only question is to distinguish between the two cases. For a qualitative argument, recall the two-capacitor model of Fig. 2.7, which consists of the oxide capacitance in series with the semiconductor (or quantum) capacitance. According to this model, a change in gate voltage produces a change in surface potential (or equivalently of $\varepsilon_1(0)/q$)

$$\delta\varepsilon_1 = \frac{-q\,\delta V_G}{1+C_S/C_{ox}}, \tag{3.44}$$

so when $C_S \gg C_{ox}$, the MOS limit occurs and the gate voltage has a small effect on the potential at the top of the barrier, while for $C_S \ll C_{ox}$, the bipolar limit occurs and the potential at the top of the barrier is controlled by the gate voltage. Figure 3.13 illustrates this by plotting $N(V_D)/N(V_D = 0)$ with C_S/C_{ox} as a parameter. For a conventional MOS capacitor operating above threshold, the semiconductor (or inversion layer capacitance) is typically much larger than C_{ox}, so the assumption that the charge is fixed by the gate voltage is a good one. As oxide thicknesses decreases, however, this assumption must be questioned. In addition, recall from Chapter 2 (Sec. 2.3) that the semiconductor capacitance is proportional to the density of states at the Fermi level, which is proportional to the effective mass. As new channel materials with small effective masses are explored, the effect of the semiconductor capacitance has to be carefully considered.

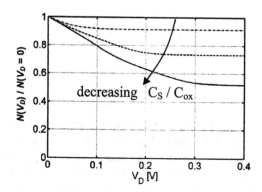

Fig. 3.13   Charge at the top of the barrier vs. drain voltage for a fixed gate voltage with the ratio, C_S/C_{ox}, as a parameter.

Having discussed charge control more carefully, we now ask how to compute the current. In the Natori model, we assume a Fermi level, a drain voltage, and then assume that

$$n_S = \frac{C_{ox}}{q}\left(V_G - V_T\right) \tag{3.45}$$

after which we can solve Eqn. (3.43a) for $\varepsilon_1(0)$. Knowing $\varepsilon_1(0)$, we can evaluate the drain current from $I_D = I^+ - I^-$ where I^+ and I^- are obtained from Eqns. (3.33e) and (3.33f). The assumption of operation in the MOS limit can be removed by simply solving the nonlinear equation, Eqn. (3.43a) and (3.43b) for $\varepsilon_1(0)$ and n_S, and then using (3.33e) and (3.33f) for the current.

3.5.2 Two Dimensional Electrostatics

As discussed in Chapter 2, two-dimensional electrostatics are an important factor for short channel MOSFETs, so we seek a way to extend our analytical model to include 2D electrostatics. As indicated in Fig. 3.14, the potential at the top of the barrier is controlled, in general, by all three electrodes. We solve for the potential at the top of the barrier by superposition. First, we ignore the mobile charge to calculate the Laplace solution from voltage division among the capacitors as

$$U_L = -q\left(\alpha_G V_G + \alpha_D V_D + \alpha_S V_S\right), \tag{3.46}$$

where

$$\alpha_G = \frac{C_G}{C_\Sigma} \quad \alpha_D = \frac{C_D}{C_\Sigma} \quad \alpha_S = \frac{C_S}{C_\Sigma}, \tag{3.47}$$

and

$$C_\Sigma = C_G + C_D + C_S. \tag{3.48}$$

Note that C_G, C_S, and C_D refer to three capacitors that describe the electrostatics control of the three electrodes. The first one is just $C_G = C_{ox}$, and C_D is chosen to reproduce the measured DIBL or output conductance of the transistor. (Do not confuse C_S with the semiconductor capacitance, which is included implicitly in this model but which does not appear explicitly.)

The next step is to ground the three terminals and compute the potential at the top of the barrier due to the mobile charge as

$$U_P = \frac{q^2}{C_\Sigma} n_S. \tag{3.49}$$

The complete solution,

$$\varepsilon_1(0) = U_L + U_P = -q\left(\alpha_G V_G + \alpha_D V_D + \alpha_S V_S\right) + U_C n_S \tag{3.50}$$

is the sum of the Laplace solution and the charging energy and can be recognized as a generalization of Eqn. (3.43a). In practice, E_F is set to give the correct off-current, α_G, to get the correct gate control, and α_D to produce the correct DIBL. Fitting procedures are described in [3.16].

In summary, the procedure is this. First, assume an energy for the top of the barrier, $\varepsilon(0)$. Next, use eqn. (3.43b) and the known Fermi levels for the source and drain to compute the charge at the top of the barrier, n_S. Finally, use eqn. (3.50) to compute an improved estimate for the energy at the top of the barrier. Repeat this process until it converges.

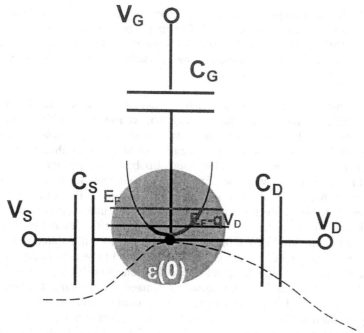

Fig. 3.14   A simple capacitor model for two-dimensional electrostatics.

3.6 Discussion

The simple treatment outlined in this chapter is complicated in practice when multiple subbands are occupied [3.14]. As discussed in Sec. 1.3, the lowest subband energy (a so-called unprimed subband) in (100) silicon, is determined by the longitudinal effective mass, and the mass in the plane of transport is the transverse effective mass. There are two such ellipsoids in (100) silicon, so the lowest subband is doubly degenerate. When computing carrier densities and currents, the appropriate valley degeneracies and effective masses must be used. For example, when computing carrier densities, one should use the density-of-states effective mass, so for the unprimed subbands,

$$m_d^* = 2m_t \tag{3.51a}$$

in N_{2D}. On the other hand, when computing the average velocity, υ_T, one should use the conductivity effective mass,

$$m_c^* = m_t. \tag{3.51b}$$

Careful bookkeeping is necessary, especially for the primed bands [3.14] and for arbitrary orientations. Different authors also define these quantities differently. Finally, note also that we have assumed simple, ellipsoidal energy bands, but the extended model described in Sec. 3.5 can readily handle arbitrary bandstructures [3.16].

One might question our use of a semiclassical model for 10 nm scale MOSFETs, for which strong quantum confinement effects might be expected. Quantum confinement is indeed a strong effect in nanoscale MOSFETs, and it leads to a significant increase in threshold voltage. Quantum mechanical tunneling can also be important, especially in the ultra-thin oxides now in use. Eventually, direct tunneling from the source to the drain sets an ultimate scaling limit. Figure 1.10b, which shows the energy-resolved electron density in a MOSFET under on-state conditions, shows strong carrier tunneling under the source to channel barrier. The effect on the MOSFET's I_{DS} vs. V_{DS} characteristic is illustrated in Fig. 3.15. The increase in off-current is readily understood to be a consequence of the source to drain tunneling, but notice that tunneling lowers the on-current. The reason is that MOS electrostatics demand that the same total charge be present at the top of the barrier in a classical or quantum model. In the quantum models, a significant fraction of this charge has to tunnel through the barrier, which impedes the current [3.5]. Even at the 10 nm scale MOSFETs operate as essentially classical devices.

a)

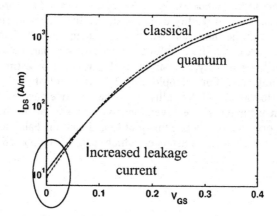

b)

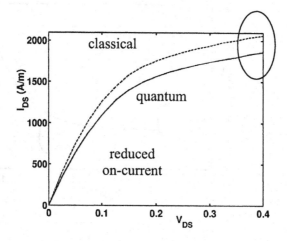

Figure 3.15   A comparison of the I-V characteristics of a 10-nm double gate MOSFET
from both a semiclassical and quantum treatment. Quantum confinement
is treated in both cases. (a) Log I_D and (b) linear I_D. (Reproduced with
permission from [1.7])

Our simple semiclassical treatment provides insights that are useful in interpreting experimental data or comprehensive simulations, and it often provides adequate estimates of the ballistic current that can be used to assess the performance of real devices.   Given the appropriate device parameters, one can readily compute the ballistic characteristics from Eqn. (3.35) or with the more sophisticated treatment described in Sec. 3.5. For quick estimates, it's often useful to know how key parameters vary with the inversion layer density, a quantity that can often be estimated reasonably well for a device. For example, Fig. 3.16 is a plot of the ballistic injection velocity (left axis) and the ballistic channel conductance as a function of n_S. Given an estimate of the inversion layer density at the beginning of the channel, one can readily use this plot to estimate the ballistic on-current and the ballistic channel resistance, which can be compared against actual devices.

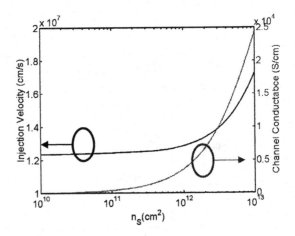

Fig. 3.16   Ballistic injection velocity (left) and the ballistic channel conductance (right axis) vs. the inversion layer density at the beginning of the channel, $n_S(0)$. One-subband occupation at room temperature is assumed and the results were obtained from Eqns. (3.36) and (3.39).

3.7 Summary

This chapter presented a thorough discussion of the ballistic MOSFET. In practice, real MOSFETs suffer from surface roughness and other scattering, which limits their performance to about 50% of the ballistic limit. Nevertheless, an understanding of the ballistic MOSFET provides a good starting point for understanding small MOSFETs – even when scattering is present. Moreover, transistors made from materials that have high mobility (low scattering) can, indeed, operate near the ballistic limit. Two examples are III-V high electron mobility transistors (HEMTs) [3.17] and the carbon nanotube transistors that we will discuss in Chapter 5.

Chapter 3 References

[3.1]   B. Doris, M. Ieong, T. Kanarsky, Y. Zhang, R.A. Roy, O. Dokumaci, Z. Ren, F-F
        Jamin, L. Shi, W. Natzle, H-J Huang, J. Mezzapelle, A. Mocuta, S. Womack, M.
        Gribelyuk, E. C. Jones, R. J. Miller, H-S P. Wong, and W. Haensch, "Extreme
        Scaling with Ultra-Thin Si Channel MOSFETs," Tech. Dig., IEEE   Electron
        Devices Mtg., pp. 267-270, Washington, Dec. 2002.

[3.2]   S. Datta, *Electronic Transport in Mesoscopic Systems,* Cambridge University Press,
        Cambridge, UK, 1997.

[3.3]   M.S. Lundstrom and Z. Ren, "Essential Physics of Carrier Transport in Nanoscale
        MOSFETs," *IEEE Trans. Electron Dev.*, **49**, pp. 133-141 , Jan. 2002.

[3.4]   E.O. Johnson, "The insulated-gate field-effect transistor - a bipolar transistor in
        disguise," *RCA Review*, **34**, pp. 80-94, 1973.

[3.5]   Z. Ren, R. Venugopal, S. Datta,, M.S. Lundstrom, D. Jovanovic, and J.G. Fossum,
        "The ballistic nanotransistor: A simulation study," IEDM Tech. Digest, pp. 715-
        718, Dec. 10-13, 2000.

[3.6]   Z. Ren, R. Venugopal, S. Datta, and M.S. Lundstrom, "Examination of design and
        manufacturing issues in a 10 nm Double Gate MOSFET using Nonequilibrium
        Green's Function Simulation," IEDM Tech. Digest, Washington, Dec. 3-5, 2001.

[3.7]   Y. Naveh and K.K. Likharev, "Modeling of 10-nm-scale ballistic MOSFET's,"
        IEEE Electron Dev. Lett., **21**, pp. 242-244, 2000.

[3.8]   Konstantin Likharev, "Electronics Below 10nm" in *Nano and Giga Challenges in
        Microelectronics*, ed. by A. Korkin, Elsevier, 2003.

[3.9]   J.-H. Rhew, Zhibin Ren, and Mark Lundstrom, "Numerical study of a ballistic
        MOSFET," S*olid-State Electronics,* **46**, No. 11, pp. 1899 – 1906, 2002.

[3.10] C. G. Sodini, P.-K. Ko, and J.L. Moll, "The Effect of High Fields on MOS Device
        and Circuit Performance," *IEEE Trans. Electron Dev.*, **31**, p. 1386, 1984.

[3.11] K. Natori, "Ballistic metal-oxide-semiconductor field effect transistor, *J. Appl.
        Phys.*, **76**, pp. 4879-4890, 1994.

[3.12] K. Natori, "Scaling limit of the MOS transistor – A Ballistic MOSFET, *IEICE
        Trans. Electron.*, **E84-C,** pp. 1029-1036, 2001.

[3.13] S. Datta, F. Assad, and M.S. Lundstrom, "The Si MOSFET from a transmission
        viewpoint," *Superlattices and Microstructures*, **23**, pp. 771-780, 1998.

[3.14] F. Assad, Z. Ren, D. Vasileska, S. Datta, and M.S. Lundstrom, "On the
        performance limits for Si MOSFET's: A theoretical study," *IEEE Trans. Electron
        Dev.*, **47**, pp. 232-240, 2000.

[3.15] R.F. Pierret, *Fundamentals of Semiconductor Devices*, Addison-Wesley, Reading,
        MA, 1996.

[3.16] Anisur Rahman, Jing Guo, Supriyo Datta, and Mark Lundstrom, "Theory of
        Ballistic Nanotransistors," *IEEE Trans. Electron. Dev.*, **50**, pp. 1853-1864, 2003.

[3.17] Jing Wang and Mark Lundstrom, "Ballistic Transport in High Electron Mobility
        Transistors,' *IEEE Trans. Electron Dev.*, **50**, pp. 1604-1609, 2003.

Chapter 4: Scattering Theory of the MOSFET

4.1 Introduction

In practice, MOSFETs operate below the ballistic limit because carriers scatter within the device. Some carriers injected from the source into the channel backscatter and return to the source. To determine how far below the ballistic limit a device operates, we can plot the ballistic *I-V* characteristic for the device as described in Chapter 3 (knowing the device's oxide thickness and power supply and threshold voltages) and then compare the results to the measured data. Typical data for circa 2000 technology with $L_{eff} \approx 100$ nm is shown in Fig. 4.1. These results show that n-channel MOSFETs deliver roughly one-half of the ballistic on-current while p-channel MOSFET deliver roughly one-third of the ballistic current. On the other hand, the measured channel conductance at low drain bias is only one tenth of the ballistic conductance. Other studies of production devices find similar results, [4.1, 4.2] although research devices that operate much closer to the ballistic limit have also been reported [4.3]. It is interesting to note that device scaling is not bringing MOSFETs closer to the ballistic limit. For the past decade or more, n-MOSFETs have operated at about or slightly below one-half of the ballistic limit. Our goal in this chapter is to develop an understanding of how scattering affects transistors and to develop a simple quantitative model that extends the ballistic model developed in the previous chapter.

Ballistic transport is simple because it is really a special type of equilibrium. Each state within the device is populated according to the equilibrium Fermi level of one of the contacts. Scattering mixes things up, and it is no longer possible to determine which contact populated a particular state. In general, this is a complicated transport problem, but we seek a simple, physical understanding.

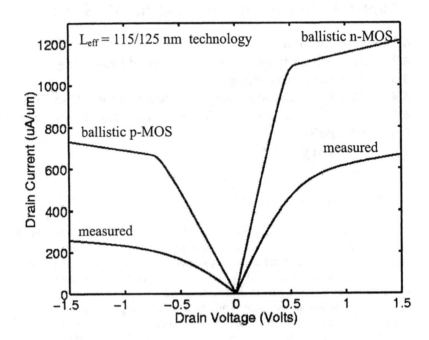

Figure 4.1   Comparison of measured I-V characteristics of p- and n-channel MOSFETs and the corresponding ballistic limit characteristics. The measured series resistance has been added to the ballistic characteristics. (From D. Rumsey, MSEE Thesis, Purdue University, December 2001.)

The channel transmission coefficient, T, provides a simple way to describe a MOSFET in the presence of scattering. Figure 4.2 defines the channel transmission coefficient. Under high drain bias in the ballistic MOSFET, only positive k-states at the top of the barrier are occupied, but in the presence of scattering, only a fraction, T_{SD}, of the current injected into the channel transmits across and exits through the drain. A fraction, $(1-T_{SD})$ backscatters and occupies negative k-states. Similarly, for low drain bias, a fraction, T_{DS}, of the flux injected into the drain transmits to the top of the barrier and adds to the population of the negative k-states. As we saw in Sec. 1.8, $T_{SD}(E) = T_{DS}(E)$ in the absence of inelastic scattering, but in general, $T_{SD} \neq T_{DS}$. We will assume, however, that the two are equal and will explain why this assumption works rather well in practice. Our objective, simply stated, is to extend Natori's theory of the ballistic MOSFET, Eqn. (3.32),

$$I_D = W \, Q(0) \, \vartheta_T \left[\frac{1 - \dfrac{F_{1/2}(\eta_F - U_D)}{F_{1/2}(\eta_F)}}{1 + \dfrac{F_0(\eta_F - U_D)}{F_0(\eta_F)}} \right]. \tag{4.1}$$

to include the effects of scattering. We might expect that current would be proportional to the channel transmission coefficient, T. We will see that it is for low drain bias but that the relationship is more involved under high drain bias.

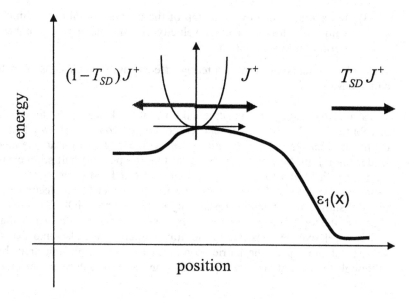

Figure 4.2   Illustration of how backscattering populates negative k-states at the top of the barrier and reduces the current transmitted to the drain.

4.2 MOSFET Physics in the Presence of Scattering

Before we develop simple models for MOSFETs in the presence of scattering, we examine the essential physics using numerical simulations based on the nonequilibrium Green's function approach, which provides a quantum mechanical description of transport [4.4, 4.5, 4.6, 4.7]. The device simulated is a double gate MOSFET because the simple geometry facilitates the numerical analysis, but the overall conclusions should apply more generally. For the ballistic MOSFET, the simulations discussed in Sec. 3.2 established that:

1) the carrier distribution function at the top of the source-channel barrier consists of two thermal equilibrium halves, one injected from the source and the other from the drain,

2) for an electrostatically well-designed MOSFET, the total carrier density at the top of the barrier is maintained at an approximately constant value, and

3) the average velocity at the top of the barrier saturates at a limiting value, which is the average velocity of a thermal equilibrium hemi-Fermi-Dirac distribution.

We seek to understand how scattering affects these conclusions from the ballistic MOSFET.

Scattering mixes the positive and negative k-states, so the carrier distribution at the top of the barrier no longer consists of two thermal equilibrium halves. (Compare Fig. 1.10b with Fig. 1.11b for an example.) Under high drain bias, however, we find that the positive half still consists of a near thermal equilibrium population injected from the source. As expected, we also find that scattering does not affect MOS electrostatics; Fig. 4.3 shows that in an electrostatically well-designed MOSFET, the total charge at the top of the barrier continues to be controlled by the gate voltage, and it is relatively insensitive to the drain voltage. We also find that the velocity at the top of the barrier continues to saturate at high drain bias (although the saturation is not quite as strong and the saturated velocity is lower).

Figure 4.4 provides some additional insight into how scattering affects a MOSFET. This plot shows the simulated on current as a function of the number of scattering sites in the simulation. For this simulation there were 45 finite difference nodes between source and drain. The first scatterer was placed at the drain, then with increasing numbers, the scattering progressively moves towards the source. Both elastic scattering and inelastic scattering in a 2D MOSFET channel are expected to relax the longitudinal energy, as discussed in more detail in 4.5. Figure 4.4 shows that when scattering is localized to the drain region, it has little effect on the on-current, but the effect of scattering increases as scattering takes place closer to the source. Although the physical model for scattering in these numerical simulations is a simple one (see [4.7] for a discussion), more rigorous simulations confirm that scattering near the drain has less influence on I_D than scattering near the source [4.8].

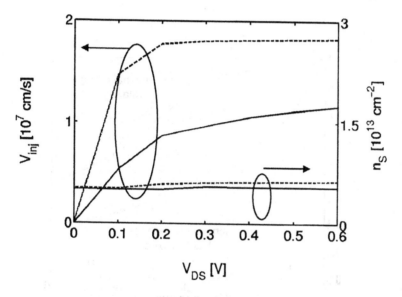

Figure 4.3   Comparison of the average carrier velocity at the top of the barrier (left) and the inversion layer density (right) when scattering is present (solid lines) and absent (dashed lines). The device is a double gate MOSFET and the results were obtained from a nonequilibrium Green's function simulation. (Reproduced with permission from [4.6])

In summary, numerical simulations in the presence of scattering show that:

1)   for an electrostatically well-designed MOSFET, the total carrier density at the top of the barrier is maintained at an approximately constant value, as it is for the ballistic MOSFET, and

2)   for a fixed gate voltage, the average velocity at the top of the barrier saturates with increasing drain voltage at a limiting value that is below the average velocity of a thermal equilibrium hemi Fermi-Dirac distribution.

3)   scattering, which controls the magnitude of the saturated velocity, is most important when it occurs near the source end of the channel.

We will use these observations to develop a scattering theory of the MOSFET in terms of a channel transmission coefficient, T, where $T < 1$ in the presence of scattering. We will use this simple theory to establish the limiting source velocity in the presence of scattering and to understand why scattering near the source is more important than near the drain.

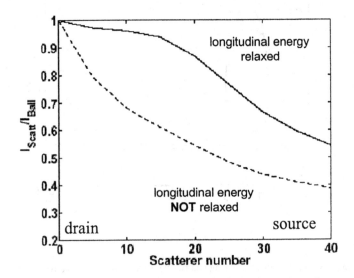

Figure 4.4   Illustration of how carrier scattering affects the on current of a 10 nm channel length double gate MOSFET. The channel is divided into 45 slabs, and scatterer 1 is placed in the slab at the drain end. With increasing number, scattering is progressively added towards the source. With 45 scatterers present, scattering takes place uniformly from source to drain. Note that these simulations were done non-selfconsistently. When 2D electrostatics are not well-controlled, scattering near the drain can play an important role. (Reproduced with permission from [4.7].)

4.3 The Scattering Model

Before we discuss the physics of scattering, we generalize Natori's ballistic theory of the MOSFET as described in Eqn. (3.4) to include scattering in terms of a phenomenological channel transmission coefficient, T. Because scattering complicates things, the derivation is not as "clean" as for the ballistic case. In addition to assuming a channel transmission coefficient whose calculation will be discussed later, the derivation will require some additional assumptions that are guided by physical reasoning.

To evaluate the drain current, we begin with

$$I_D = I^+ - I^- = I^+\left(1 - I^-/I^+\right) = W q n_S^+ v^+\left(1 - I^-/I^+\right),$$ (4.2)

where the carrier densities and velocities are evaluated at the top of the barrier. We also have

$$n_S^+(0) + n_S^-(0) = Q(0)/q,$$ (4.3a)

which can be solved for

$$q n_S^+(0) = \frac{Q(0)}{\left[1 + n_S^-/n_S^+\right]}.$$ (4.3b)

Equation (4.3b) may be inserted into Eqn. (4.2) to find the general expression for the drain current as

$$I_D = W Q(0) v^+ \left[\frac{1 - I^-/I^+}{1 + n_S^-/n_S^+}\right],$$ (4.4)

which is just Eqn. (3.4) for the ballistic case. So far, we have made no assumptions; to make progress we now need to make some.

We begin by <u>assuming</u> that the positive directed moments are given by their ballistic values as in Chapter 3,

$$I^+ = I_b^+ = W q \left(\frac{N_{2D}}{2}\right) v_T F_{1/2}(\eta_F)$$ (4.5a)

and

$$n_S^+(0) = n_b^+(0) = \left(\frac{N_{2D}}{2}\right) F_0(\eta_F),$$ (4.5b)

where the subscript, *b*, is a reminder that we are using the ballistic expressions of Chapter 3. The use of equilibrium quantities for the positive-directed moments is analogous to the quasi-equilibrium assumption for forward biased junctions, which leads to the well-known "Law of the

Junction" [4.9]. Our next task is to determine the negative-directed flux and carrier density.

The negative flux consists of two components, one injected from the drain and another that consists of the fraction of the positive flux injected from the source that backscatters. In the case of ballistic transport, we saw in Chapter 3 that the contribution from the drain (the only contribution in the ballistic case) is

$$I_b^- = W q \left(\frac{N_{2D}}{2} \right) \upsilon_T F_{1/2} \left(\eta_F - U_D \right)$$
(4.6a)

and

$$n_b^-(0) = \left(\frac{N_{2D}}{2} \right) F_0 \left(\eta_F - U_D \right).$$
(4.6b)

When scattering is present, the contribution to the negative k-states from drain injection will be reduced by a drain-to-source transmission coefficient. The contribution to the negative k-states from source injection will involve a channel backscattering coefficient, $\left(1 - T_{SD}\right)$. We can, therefore, write the negative current at the top of the barrier as

$$I^- = (1 - T_{SD}) I^+ + T_{DS} I_b^- ,$$
(4.7)

where the first term is the contribution from backscattering of the injected current and the second term is the contribution from the drain.

To find the carrier density in the negative k-states, we divide each flux by its appropriate velocity to find

$$n_S^- = \frac{I^-}{qW\upsilon^-} = (1 - T_{SD}) \frac{I^+}{qW\upsilon_S^-} + T_{DS} \frac{I_b^-}{qW\upsilon_D^-} ,$$
(4.8)

where υ^- is the average velocity of the negative-directed current, υ_S^- is the average velocity of the portion of the negative–directed flux that arises from the backscattering of source-injected carriers, and υ_D^- is the average velocity of the portion of the negative-directed flux due to injection from the drain.

We now make some additional assumptions. First, we assume that

$$\bar{\upsilon_D} = \upsilon_T \frac{F_{1/2}(\eta_F - U_D)}{F_0(\eta_F - U_D)},$$ (4.9a)

that is, we assume that it is equal to its value in equilibrium. We also assume that

$$\bar{\upsilon_S} = \upsilon_T \frac{F_{1/2}(\eta_F)}{F_0(\eta_F)},$$ (4.9b)

which means that we assume that the average velocity of the backscattered source-injected carriers is equal to that of the positive-velocity source-injected carriers. Finally, we assume that

$$T_{SD} = T_{DS} = T.$$ (4.9c)

With these assumptions, Eqn. (4.8) becomes

$$n_S^- = (1-T)n_S^+ + Tn_b^-$$ (4.10)

so

$$n_S^- / n_S^+ = (1-T) + T\left[F_0(\eta_F - U_D)/F_0(\eta_F)\right].$$ (4.11a)

Using assumption (4.9c) in Eqn. (4.7), we also find

$$I^- / I^+ = (1-T) + T\left[F_{1/2}(\eta_F - U_D)/F_{1/2}(\eta_F)\right].$$ (4.11b)

Equations (4.11a) and Eqn. (4.11b) can be inserted into Eqn. (4.4) to find

$$I_D = W\, Q(0)\left(\frac{T}{2-T}\right)\vartheta_T \left[\frac{1 - \dfrac{F_{1/2}(\eta_F - U_D)}{F_{1/2}(\eta_F)}}{1 + \left(\dfrac{T}{2-T}\right)\dfrac{F_0(\eta_F - U_D)}{F_0(\eta_F)}}\right],$$ (4.12)

which is the desired result.

Equation (4.12) generalizes the expression for the ballistic MOSFET, Eqn. (4.1), to the case where scattering is present. Note that the current is not simply proportional to the transmission coefficient - the relation is more involved. To arrive at this simple expression, we needed to make a number of assumptions beyond those in Natori's ballistic model. The key additional assumptions are:

1) that the positive flux injected into the channel continues to be the thermal equilibrium flux injected over the barrier from the source,

2) that the average velocity of those negative-velocity carriers at the top of the barrier that are due to backscattering of the source-injected stream is equal to that of the positive-velocity source-injected stream,

3) that the average velocity of those negative-velocity carriers at the top of the barrier that are due to the drain-injected stream is equal to that of the ballistic drain-injected stream,

4) that the average transmission coefficient from the source to drain, T_{SD}, is equal to that from the drain to the source, T_{DS}.

These assumptions seem to capture the essence of the problem, but they are hard to justify rigorously. Assumption 1) is commonly used for problems that involve injection over a barrier. Assumptions 2), 3), and 4) can be justified at and near equilibrium ($V_{DS} \approx 0$). For large V_{DS}, drain injection is unimportant, so assumptions 3) and 4) have little effect on the high V_{DS} drain current. For high drain bias, we justify assumption 2) by assuming that elastic, isotropic scattering of the source-injected stream dominates at the top of the barrier, so that the energy distribution of the backscattered stream is identical to that of the source-injected stream. The assumption that $T_{SD}(E) = T_{DS}(E)$ is rigorously true in equilibrium or in the absence of inelastic scattering (recall the discussion in Sec. 1.8). Under low drain bias, the MOSFET is near equilibrium, and the assumption is valid, and under high drain bias, only T_{SD} matters.

Equation (4.12) is a simple expression for the *I-V* characteristic in the presences of scattering, but note that it hides a great deal of complexity in the transmission coefficient, T. Comparisons with experimental results like those in Fig. 4.1 show that T is bias dependent. At low drain bias, $T \approx 0.1$ for an $L_{eff} \approx 100$ nm n-channel MOSFET (≈ 0.05 for a p-MOSFET) while at high drain bias, $T \approx 0.67$ for an n-MOSFET(≈ 0.50 for a p-MOSFET). Computing T rigorously is a difficult problem that requires rigorous numerical simulations, but the essential physics of scattering in a MOSFET can be understood simply, as discussed in the next two sections.

Before discussing the physics of scattering, it's useful to examine Eqn. (4.12) under two limiting conditions. First of all, it's clear that in the ballistic limit where $T = 1$, we recover the ballistic theory of the MOSFET. Under low drain bias, we can expand the drain voltage factor for small V_{DS} to find

$$I_D = W\,C_{ox}\left(V_G - V_T\right)\frac{T\,\vartheta_T}{2\left(k_B T_L/q\right)}\left(\frac{F_{-1/2}(\eta_F)}{F_{1/2}(\eta_F)}\right)V_D,\qquad(4.13a)$$

Where we have inserted $Q(0) = C_{ox}\left(V_{GS} - V_T\right)$. From Eqn. (4.13a), we find

$$G_{CH} = T\left[\frac{W C_{ox}\left(V_{GS} - V_T\right)\vartheta_T}{2\left(k_B T_L/q\right)}\left(\frac{F_{-1/2}(\eta_F)}{F_{1/2}(\eta_F)}\right)\right] = T G_B,\qquad(4.13b)$$

where G_B is the ballistic channel conductance as given by Eqn. (3.36). We might have guessed that the channel conductance would be T times the ballistic result.

For high drain bias, the drain voltage factor approaches unity and we find [4.11]

$$I_D(on) = W\,C_{ox}\left\{\left(\frac{T}{2-T}\right)\vartheta_T\right\}\left(V_G - V_T\right),\qquad(4.14)$$

which has a more involved dependence on T. (The dependence on T arises from the requirement of MOS electrostatics that $Q(0)$, which under high drain bias consists only of the source-injected stream and the backscattered stream, be independent of drain bias.) The average carrier velocity at the top of the barrier,

$$\langle v(0)\rangle = \left(\frac{T}{2-T}\right)\vartheta_T,\qquad(4.15)$$

is reduced by backscattering.

4.4   The Transmission Coefficient under Low Drain Bias

Under low source to drain bias, the channel of a nanoscale MOSFET can be regarded as a region of length, L_{eff}, across which carriers diffuse. (Strictly speaking, the dominance of diffusion over drift implies that the potential drop across the channel is less than $k_B T_L/q$.) For these conditions, the channel transmission coefficient is readily evaluated.

Consider a flux of carriers in the positive direction, J^+ and a corresponding flux of carriers in the negative direction, J^-. The positive stream is reduced by scattering out but increased when the negative stream scatters in. The result is described by McKelvey's equations [4.14]

$$\frac{dJ^+}{dx} = -\frac{J^+}{\lambda_o} + \frac{J^-}{\lambda_o}, \tag{4.16a}$$

$$\frac{dJ^-}{dx} = -\frac{J^+}{\lambda_o} + \frac{J^-}{\lambda_0}, \tag{4.16b}$$

where λ_o is the near-equilibrium mean-free-path for backscattering (assumed constant) and dx/λ_o is the probability of scattering in a thickness, dx. (We use the near-equilibrium mean-free-path because we assume that an equilibrium flux is injected into the slab and that there is no electric field within the slab.) The sign is correct for the second equation (note that J^- is a positive quantity that points in the $-x$ direction). Using the net flux, $J = \left(J^+ - J^-\right)$, to eliminate J^- in Eqn. (4.16), we find

$$\frac{dJ^+}{dx} = -\frac{J}{\lambda_o}, \tag{4.17}$$

which can be integrated to find

$$J^+(x) = J^+(0) - J\left(\frac{x}{\lambda_o}\right). \tag{4.18}$$

Now consider a slab of length, L, which represents the channel under low drain bias. As shown in Fig. 4.5, we inject a flux of positive velocity carriers into the slab at $x = 0$, and a flux $T J^+(0)$ emerges from the right where we assume a perfectly absorbing contact so that no negative flux enters the slab from the right. Within the slab, scattering converts part of the injected positive stream to the negative stream, so we have both positive and negative fluxes,

Our goal is to evaluate T. At $x = L$ we have only a positive flux, so $J = J^+(L) = T J^+(0)$. Using this expression in Eqn. (4.18), we find

$$T = \frac{J^+(L)}{J^+(0)} = \frac{\lambda_o}{\lambda_o + L}. \qquad (4.19)$$

As expected, the channel transmission coefficient approaches zero when $L \gg \lambda_o$ and one when $L \ll \lambda_o$. In general, the mean-free-path depends on the energy of the carriers. If we inject an equilibrium flux into the device, then the appropriate mean-free-path is λ_o

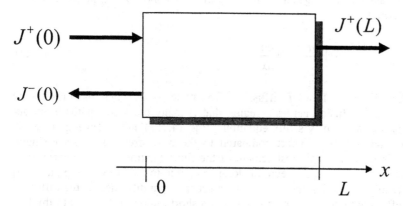

Figure 4.5   Carrier transport across a field-free semiconductor slab. A flux is injected at $x = 0$ and absorbed at $x = L$.

Using Eqn. (4.19) in the linear region current, Eqn. (4.13a), we find

$$I_D = \left(\frac{W}{L+\lambda_o}\right) C_{ox}(V_G - V_T) \frac{\lambda_o \upsilon_T}{2(k_B T_L/q)} \left(\frac{F_{-1/2}(\eta_F)}{F_{1/2}(\eta_F)}\right) V_D \qquad (4.20a)$$

$$I_D = \left(\frac{W}{L+\lambda_o}\right) C_{ox}(V_G - V_T) \frac{\upsilon_T \lambda_o}{(2k_B T_L/q)} V_D, \qquad (4.20b)$$

where we have assumed nondegenerate carrier statistics in the second expression. To apply Eqn. (4.20) in practice, we need to estimate the mean-free-path for backscattering.

To relate the mean-free-path to a familiar macroscopic quantity, we add the flux equations, Eqns. (4.16a) and (4.16b), to obtain

$$\frac{d\left(J^{+}+J^{-}\right)}{dx} = -\frac{2}{\lambda_o}\left(J^{+}-J^{-}\right) = -\frac{2J}{\lambda_o}, \tag{4.21}$$

which can be solved for the net flux

$$J = -\left(\frac{\lambda_o}{2}\right)\frac{d\left(J^{+}+J^{-}\right)}{dx} = -\left(\frac{\lambda_o \upsilon_T}{2}\right)\frac{d\left(J^{+}/\upsilon_T + J^{-}/\upsilon_T\right)}{dx}. \tag{4.22}$$

If we assume nondegenerate, near-equilibrium conditions, Eqn. (4.22) can be expressed as

$$J = -\left(\frac{\lambda_o \upsilon_T}{2}\right)\frac{dn}{dx} = -D\frac{dn}{dx}, \tag{4.23}$$

which is Fick's Law of diffusion. One typically assumes that Fick's Law holds for diffusion across a region that is many mean-free-paths long, but note that McKelvey's flux equations, Eqs. (4.16a) and (4.16b) apply when the slab is long or short compared to the mean-free-path. Surprisingly, Fick's Law, a simple restatement of the flux equations, is not restricted to situations in which the region is long compared to the mean-free-path [4.15]. One might question the use of a macroscopic quantity such as the diffusion coefficient to describe transport across a short region. It is better to think in terms of the mean-free-path for backscattering, λ_o. The probability of backscattering in a length, dx, is simply dx/λ_o, which is well defined – even in a short region.

Knowing the diffusion coefficient or mobility in a bulk semiconductor, we can estimate the near-equilibrium mean-free-path from

$$D \equiv \lambda_o \upsilon_T/2 = \left(k_B T_L/q\right)\mu_{eff}. \tag{4.24}$$

The connection to the carrier mobility arises because we have assumed near-equilibrium, nondegenerate conditions for which an Einstein relation applies.

Equation (4.24) provides us with a simple way to estimate the mean-free-path if the mobility is known (subject, however, to the constraint of near-equilibrium, nondegenerate conditions). It also allows us to write the linear region drain current, Eqn. (4.20b), in a more familiar form

$$I_D = \left(\frac{W}{L + \lambda_o} \right) \mu_{eff} \, C_{ox} \left(V_G - V_T \right) V_D. \qquad (4.25)$$

Equation (4.25) is identical to the traditional textbook expression [4.9] except that the channel length, L, is replaced by $L + \lambda_o$. As the channel length becomes much smaller than the mean-free-path, the channel conductance approaches the ballistic limit.

When parasitic series resistance is eliminated, the measured linear region current is observed to be about 0.1 of the ballistic current for 100nm generation n-MOSFETs. From Eqn. (4.13a), this ratio is T, so for a $L \approx$ 100nm channel length, we infer that $\lambda_o \approx 10$ nm. According to Eqn. (4.24), this corresponds to a mobility of approximately 250 cm^2/V-s, which is roughly the expected electron mobility for a 100nm MOSFET. Within the accuracy of our assumptions, we conclude that this simple theory roughly accounts for scattering under low drain voltages.

4.5   The Transmission Coefficient under High Drain Bias

Under high drain bias it is considerably more difficult to develop a simple expression for the channel transmission coefficient. For such conditions, carrier transport is far from equilibrium. Well-known off-equilibrium transport effects such as velocity overshoot occur and simple analytical treatments are hard to come by [4.12]. Nevertheless, we seek a "back-of-the-envelope" analytical expression for T that is consistent with experiment and provides physical insight.

Figure 4.6 is a sketch of the conduction subband profile vs. position along the channel of a small MOSFET under high drain and gate bias. The channel consists of two portions, a low field region near the source, which is under strong gate control, and a high field region near the drain. The transition between the low and high field regions is often rather sharp. Transport in the channel of a MOSFET, therefore, is similar to transport in a bipolar transistor. Carriers diffuse across the low-field portion of the channel (which is analogous to the base) after which they are collected by the high-field portion of the channel, which is analogous to the collector. Following this reasoning, we postulate that the transmission coefficient under high drain bias is given by

$$T = \frac{\lambda_o}{\lambda_o + \ell}, \qquad (4.26)$$

which is analogous to Eqn. (4.19). In Eqn. (4.26), ℓ is a critical length, some fraction of the channel length that has to be specified. We have seen that for low drain bias, $\ell \approx L$. Since diffusion dominates drift when the potential drop across a region is less than $k_B T_L/q$, we are tempted to identify ℓ as the distance over which the first $k_B T_L/q$ of potential drops. If we did so, our result would be analogous to the well-known Bethe condition for thermionic emission in a metal-semiconductor diode [4.13]. Our use of the near-equilibrium mean-free-path is justified by the fact that Eqn. (4.26) describes transport across the low-field portion of the channel, when the injected equilibrium flux gains little energy.

Equation (4.26) can be mathematically derived from McKelvey's flux theory (see [4.12, 4.14, 4.15]), but the flux equations are based on a near equilibrium assumption which makes the derivation little better than the plausibility argument we have given. The physical reasoning that leads to Eqn. (4.26) is, however, consistent with numerical experiments by Monte Carlo simulation. These simulations show what happens when carriers are injected into a high field region. If they penetrate more than a short distance into the high-field region, then even if they do backscatter, they are unlikely to re-enter the injecting contact [4.16, 4.11]. Our task, therefore, is to identify the distance, ℓ, beyond which if a carrier scatters it will be unlikely to return to the source.

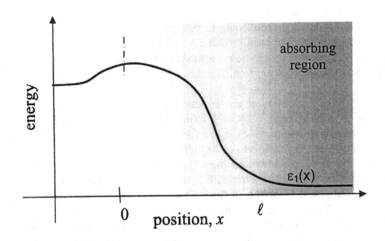

Figure 4.6   Illustration of the conduction subband profile vs. position under high drain bias. For a well-designed MOSFET, the channel is typically divided into a low field region near the source and a high field region near the drain.

Figure 4.7a illustrates a scattering event for an electron with energy E_i (a kinetic energy of $(E_i - E_b)$) that is injected in the channel and scatters first at location $x = x_l$. The electron enters the channel with some momentum, p_0, in the x-y plane, as shown in Fig. 4.7b. The longitudinal electric field in the channel accelerates the electron so that its x-directed momentum increases. At x_l, where it scatters first, its momentum is p_l, where

$$\frac{p_1^2}{2m^*} = (E_i - E_b) + qV(x_1).$$

(4.27)

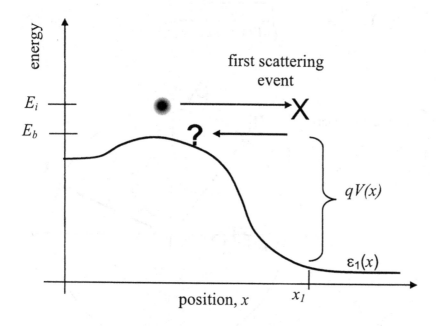

Figure 4.7a   Illustration of an electron injected into the channel with an energy, E_i, that undergoes its first scattering event at $x = x_l$. We seek to compute the probability that it returns to the source. It is important to remember that the electron is also free to move in the y-direction.

We assume that the electron scatters isotropically and elastically to momentum p_1'. After scattering, an electron can surmount the potential barrier, $qV(x_1)$, and exit through the source <u>only</u> if its velocity is in the negative x-direction and if $(1/2)m^*v_x^2 > qV(x_1)$. The shaded region in Fig. 4.7b is the only fraction that contributes to channel backscattering as we have defined it, and one can readily show that to surmount the barrier

$$\theta < \cos^{-1}\left(\sqrt{\frac{qV(x_1)}{qV(x_1)+(E_i - E_b)}}\right). \tag{4.28}$$

The probability that an electron which scatters first at $x = x_1$ returns to the source is

$$P(x_1) = \frac{2\theta}{2\pi} = \frac{1}{\pi}\cos^{-1}\sqrt{\frac{qV(x_1)}{qV(x_1)+(E_i - E_b)}} \tag{4.29}$$

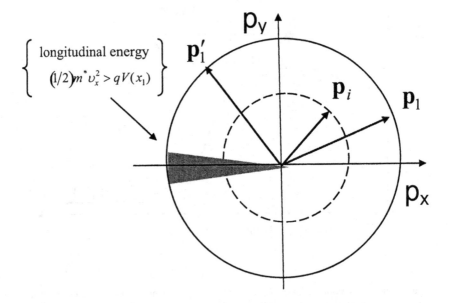

Figure 4.7b Illustration of the scattering event of Fig. 4.6a in momentum space. Carriers that backscatter into the shaded region have sufficient longitudinal energy to surmount the barrier and return to the source.

Figure 4.8 is a plot of Eqn. (4.29) vs. potential drop. It shows that when electrons gain more kinetic energy in the channel field $[qV(x_l)]$ than the kinetic energy with which they were injected into the channel, $(E_i - E_b)$, then when they scatter they are unlikely to have sufficient kinetic energy associated with the x-directed velocity to surmount the barrier and return to the source. The answer to our question of how to define the critical length in Eqn. (4.26), therefore is: the critical length for channel backscattering is the distance over which the carrier gains an energy in the channel electric field that is equal to the kinetic energy that it was injected with. Nondegenerate carriers are injected with a kinetic energy of $k_B T_L$, so the critical layer is the part of the channel over which the first $k_B T_L/q$ of potential drop occurs, which is typically a small part of the channel. Above threshold, carriers are degenerate and their kinetic energy exceeds $k_B T_L$.

Consider a typical case for a high performance transistor, $n_S \approx 10^{13}$ cm^{-2} and the Fermi level is located about 125 meV above the bottom of the subband. Power supply voltages are still several times higher than the injected kinetic energy, so the portion of the channel that contributes to backscattering to the source is a rather small fraction of the channel. This explains why T increases with V_{DS}. The increase in T is due to the reduction in the length of the low field region. There is actually more scattering under high drain bias, but it occurs near the drain end of the channel where it has less effect on T and I_D.

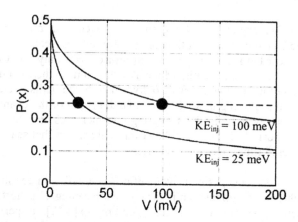

Figure 4.8   Plot of P(x) vs. V(x) from Eqn. (4.29) for two different injection kinetic energies, 25 and 100 meV. The plot shows that the transition between a high probability of backscattering to the source and a low probability occurs when the energy gained in the channel electric field equals the kinetic energy with which the carrier was injected into the channel.

4.6   Discussion

Our view of how scattering affects the drain current is based on expressing the transmission coefficient as $T = \lambda_o/(\lambda_o + \ell)$. Under low drain bias, the critical length, ℓ, is the entire channel, but under high drain bias, it is a portion of the channel near the source end. These simple arguments are supported by more rigorous numerical simulations [4.7, 4.8]. But one important factor has been overlooked – self-consistent electrostatics. Scattering near the drain does matter because when carriers scatter their density builds up which affects the self-consistent potential throughout the channel, including the critical region near the source. As channel lengths become ultrasmall, this is becoming an important factor [4.7, 4.8].

According to our scattering theory, the on current of a MOSFET is controlled by the ratio

$$B = \frac{I_D(\text{on})}{I_D(\text{on}-\text{ballistic})} = \frac{T}{2-T} = \frac{\lambda_o}{2\ell+\lambda_o}, \tag{4.30}$$

where we refer to B as the ballistic figure of merit for the device. This factor approaches one and the on-current approaches its ballistic limit when

$$\lambda_o \gg 2\ell. \tag{4.31}$$

Equation (4.31) explains why present-day devices can operate relatively close to the ballistic limit even though they are several mean-free-paths long. To deliver near-ballistic currents, the mean-free-path only has to be much longer than the critical length, which is a fraction of the channel length. Note also that the fact that a device delivers nearly the ballistic current <u>does not</u> imply that scattering is weak. Near the drain end of the channel, where the kinetic energy is high, the scattering rate increases and the mean-free-path is very short. We have seen, however, that carriers that scatter near the drain have less influence on the on current.

Mobility is a concept that is well defined in a bulk semiconductor under near equilibrium conditions; but it has no clear meaning in a short high-field region where transport is very far from equilibrium [4.12]. Experimentalists observe, however, a correlation between the on current of a small transistor and the near-equilibrium mobility [4.18, 4.19]. The near-equilibrium mobility, however, is related to the near-equilibrium mean-free-path which controls the on current of a nanoscale MOSFET. In this sense, we can say that mobility is relevant to nanoscale MOSFETs. In fact, one can show that the fractional change in current is related to the fractional change in mobility according to [4.17]

$$\frac{\Delta I_D}{I_D} = \frac{\Delta \mu_{eff}}{\mu_{eff}} (1-B).$$

(4.32)

Under low drain bias, we saw that $B \approx 0.1$, so $\Delta I_D / I_D \approx \Delta \mu_{eff} / \mu_{eff}$ as is observed. Under high drain bias, $B \approx 0.5$, so $\Delta I_D / I_D \approx 0.5 \times \Delta \mu_{eff} / \mu_{eff}$, as is also observed. New channel materials with higher mobilities are being explored as a means of increasing the on-current [4.20], but as we approach the ballistic limit, mobility will become less and less relevant.

Present day devices operate well below the ballistic limit $(T < 1)$, but it is of interest to ask how T scales as device dimensions shrink. According to Eqn. (4.26)

$$T = \frac{1}{1 + \ell / \lambda_o}.$$

(4.33)

As device dimensions shrink, the critical length, ℓ, will also shrink. At the same time, however, the increased doping and the stronger normal electric fields associated with the thinner gate oxide reduce the mean-free-path, λ_o. The result is that T is roughly constant with scaling. This expectation is confirmed by analyzing device data for the past 15 years. Devices are scaled in a way that maintains a constant transmission coefficient.

In Sec. 4.4, we showed that the scattering theory for the linear region could be written in the conventional form, except that L was replaced by $L + \lambda_o$. It turns out that the high V_{DS} current, Eqn. (4.13a), can also be expressed in a more conventional form. When $V_{DS} < V_{Dsat}$, the applied potential drops approximately linearly, so

$$\ell \approx \left(\frac{k_B T_L / q}{V_{DS}} \right) L.$$

(4.34)

If Eqn. (4.34) is used with Eqn. (4.33) in the drain current expression, Eqn. (4.14), we can express the drain current as

$$I_D = \frac{W C_{ox} \upsilon_T \, m (V_{GS} - V_T) V_{DS}}{1 + m V_{DS}},$$

(4.35)

where

$$m = \frac{\mu}{L \upsilon_T} \qquad (4.36)$$

and we have assumed nondegenerate conditions. Equation (4.35) is identical to a more conventional short channel model developed by Veeraraghavan and Fossum [4.21] except that the saturated velocity is replaced by the thermal velocity and a bulk charge term in [4.21] is missing in (4.35). Although the expressions are similar, the scattering approach and the conventional approach are based on much different physical pictures. Beginning from the velocity saturation view, we would be tempted to say that if velocity overshoot occurs, the on-current should be enhanced. In the scattering view, however, the limiting velocity is set by source injection. Velocity overshoot occurs within the channel where it can influence the self-consistent electrostatic potential. Since this affects the potential at the source, velocity overshoot influences the current indirectly.

For very long channel devices, V_{Dsat} approaches $V_{GS} - V_T$. If we substitute $V_{Dsat} = (V_{GS} - V_T)$ and assume that L is large, Eqn. (4.35) becomes

$$I_D = \mu C_{ox} \frac{W}{2L} (V_{GS} - V_T)^2, \qquad (4.37)$$

which is the expected result for a long channel transistor. We conclude that the scattering model produces results much like the conventional model in both the long and short-channel limits [4.22].

4.7   Summary

Figure 4.9 summarizes the essential physical picture that we have developed for the nanoscale MOSFET. Near thermal equilibrium conditions are maintained at the source end of the channel so that $Q_n = C_{ox}(V_{GS} - V_T)$ (two-dimensional electrostatics lowers V_T from its value in an MOS capacitor or long channel MOSFET). Since carriers are in thermal equilibrium at the source, the maximum velocity at the source end of the channel is limited to the thermal injection velocity, υ_T, which can exceed the bulk saturation velocity, 1.0 x 10^7 cm/s, by as much as 50% when the carriers are degenerate.

In practice, the average carrier velocity at the source is less than ϑ_T because of carrier backscattering. For steady-state current, the backscattering that matters the most occurs within a critical distance, ℓ, from the beginning of the channel. The length of this critical region is roughly the distance over which the potential drops by the average kinetic energy of the injected carriers. Backscattering within the critical region is related (through the mean-free-path, λ_o) to the low-field, near-equilibrium mobility of inversion layer carriers. Carriers that backscatter deeper within the channel are likely to emerge from the drain where they contribute to the DC drain current. (Scattering throughout the channel influences the device transit time, but the transit time is very short in a nanoscale transistor.) Deep within the channel, carrier transport is complex, and strong velocity overshoot occurs. Velocity overshoot lowers the carrier density near the drain, which influences the self-consistent electric field, and therefore has an indirect effect on the drain current. (As devices get very small, however, this indirect effect may become quite strong.)

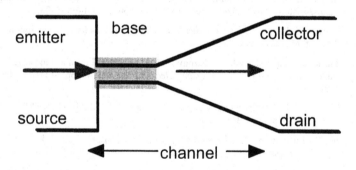

Figure 4.9   Essential physical picture of steady-state carrier transport in the nanoscale MOSFET.

Chapter 4 References

[4.1] F. Assad, Z. Ren, S. Datta, M.S. Lundstrom, and P. Bendix, "Performance limits of Si MOSFET's," *IEDM Tech. Digest*, pp. 547-549, Dec. 1999.

[4.2] A. Lochtefield and D. Antoniadis, "On experimental determination of carrier velocity in deeply scaled NMOS: How close to the thermal limit?, *IEEE Electron Dev. Lett.*, **22**, pp. 95-97, 2001.

[4.3] G. Timp, J. Bude, et al., *IEDM Tech. Digest*, pp. 55-58, 1999.

[4.4] Z. Ren, R. Venugopal, S. Datta,, M.S. Lundstrom, D. Jovanovic, and J.G. Fossum, "The ballistic nanotransistor: A simulation study," *IEDM Tech. Digest*, pp. 715-718, Dec. 10-13, 2000.

[4.5] Z. Ren, R. Venugopal, S. Datta, and M.S. Lundstrom, "Examination of design and manufacturing issues in a 10 nm Double Gate MOSFET using Nonequilibrium Green's Function Simulation," IEDM Tech. Digest, Washington, D.C., Dec. 3-5, 2001.

[4.6] Mark Lundstrom and Zhibin Ren, "Essential Physics of Carrier Transport in Nanoscale MOSFETs," *IEEE Trans. Electron Dev.*, **49**, pp. 133-141, January, 2002.

[4.7] R. Venugopal, M. Paulsson, S. Goasguen, S. Datta, and M.S. Lundstrom, "A Simple Quantum Mechanical Treatment of Scattering in Nanoscale Transistors," *J. Appl. Phys.*, **93**, pp. 5613-5625, May 1, 2003.

[4.8] A. Svizhenko, M. Anantram, and T. Govindan, "The role of scattering in nanotransistors," *IEEE Trans. Electron Dev.*, **50**, pp. 1459-1466, 2003.

[4.9] Robert F. Pierret, *Advanced Semiconductor Fundamentals*, Addision-Wesley, Reading, Massachusetts, 1987.

[4.10] Anisur Rahman and Mark Lundstrom, "A Compact Model for the Nanoscale Double Gate MOSFET," *IEEE Trans. Electron Dev.*, vol. 49, pp. 481-489, March 2002.

[4.11] M. S. Lundstrom, "Elementary scattering theory of the MOSFET," *IEEE Electron Dev. Lett.*, **18**, pp. 361-363, 1997.

[4.12] M. S. Lundstrom, *Fundamentals of Carrier Transport*, 2nd Ed., Cambridge University Press, Cambridge, UK, 2000.

[4.13] E.H. Rhoderick, *Metal-Semiconductor Contacts*, Clarendon Press, Oxford, UK, 1978.

[4.14] J.P. McKelvey, R.L. Longini, and T.P. Brody, "Alternative approach to the solution of added carrier transport problems in semiconductors, " *Phys. Rev.*, **123**, pp. 51-57, 1961.

[4.15] W. Shockley, "Diffusion and drift of minority carrier in semiconductors for comparable capture and scattering mean free paths," *Phys. Rev.*, **125**, pp. 1570-1576, 1962.

[4.16] P.J. Price, "Monte Carlo calculation of electron transport in solids," *Semiconductors and Semimetals*, **14**, pp. 249-334, 1979.

[4.17]   Mark Lundstrom, "On the Mobility Versus Drain Current Relation for a Nanoscale MOSFET," IEEE Electron Dev. Lett.,. **22**, No. 6, pp. 293-295, 2001.

[4.18]   A. Lochtefeld and D.A. Antoniadis, "Investigating the relationship between electron mobility and velocity in deeply scaled NMOS via mechanical stress," *IEEE Electron Dev. Lett.*, **22**, pp. 591-593, 2001.

[4.19]   R. Ohba and T. Mizuno, "Nonstationary electron/hole transport in Sub-0.1mm MOS devices: Correlation with mobility and low power CMOS application," *IEEE Trans. Electron Dev.*, **48**, pp. 338-343, 2001.

[4.20]   K. Rim, J.L. Hoyt, and J.F. Gibbons, "Fabrication and Analysis of Deep Submicron Strained-Si N-MOSFET's," *IEEE Trans. Electron Dev.*, **47**, pp. 1406-1415, 2000.

[4.21]   S. Veeraraghavan and J.G. Fossum, "A Physical Short-Channel Model for the Thin-Film SOI MOSFET Applicable to Device and Circuit CAD," *IEEE Trans. on Electron Dev.*, **35**, pp. 1866-1875, 1988.

[4.22]   This connection was first pointed out to me in discussions with Professor J.G. Fossum', Lixin Ge, and Keunwoo Kim at the University of Florida, March, 1999.

Chapter 5: Nanowire Field-Effect Transistors

5.1 Introduction

In a conventional MOSFET, carriers are confined in a direction normal to the channel, and free to move in two dimensions. In the previous two chapters, we discussed transistors based on such 2D carriers. It is, however, now possible to make structures that confine carriers in two dimensions, so that they are free to move only in one direction. We refer to such structures as 1D nanowires, and during the past few years, the ability to fabricate field-effect transistors from nanowires has progressed rapidly [5.1 – 5.5]. Our purpose in this chapter is to discuss the theory of nanowire field-effect transistors using two examples – the semiconductor (specifically, silicon) nanowire FET and the carbon nanotube FET. We begin with a discussion of a simple theory for the ballistic silicon nanowire FET.

5.2 Silicon Nanowire MOSFETs

The approach of Chapter 3 can be used to establish some general features of semiconductor nanowire MOSFETs. We assume a very simple geometry as shown in Fig. 5.1 – a nanowire that is coaxially gated. Instead of $C_{ins} = \kappa_{ins}\varepsilon_0/t_{ins}$ F/cm^2 as for a MOSFET, we have an insulator capacitance of

$$C_{ins} = \frac{2\pi\kappa\varepsilon_0}{\ln\left(\dfrac{2t_{ins} + t_{wire}}{t_{wire}}\right)} \qquad \text{F/cm,} \qquad (5.1)$$

where t_{wire} is the diameter of the wire.

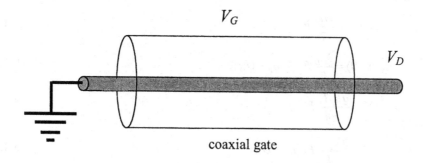

$$V_G$$

$$V_D$$

coaxial gate

Figure 5.1   The geometry of a simple, idealized coaxial gate nanowire MOSFET.

We first need to evaluate some directed moments analogous to Eqns. (3.5). Specifically, we must evaluate

$$n_L^+(0) = \frac{1}{L}\sum_{k>0} f_0(E_F) \qquad \text{cm}^{-1} \qquad (5.2a)$$

$$n_L^-(0) = \frac{1}{L}\sum_{k<0} f_0(E_F - qV_D) \qquad \text{cm}^{-1} \qquad (5.2b)$$

$$I^+ = \frac{1}{L}\sum_{k>0} q\upsilon f_0(E_F) \equiv qn_L^+(0)\upsilon^+(0) \qquad \text{A} \qquad (5.2c)$$

$$I^- = \frac{1}{L}\sum_{k<0} q\upsilon f_0(E_F - qV_D) \equiv qn_L^-(0)\upsilon^-(0) \quad \text{A} \qquad (5.2d)$$

We will work within the effective mass approximation and assume a simple, parabolic bandstructure,

$$\varepsilon_1(k) = \varepsilon_1(0) + \frac{\hbar^2 k^2}{2m^*},$$ (5.3)

where $\varepsilon_1(0)$ is the minimum of the first subband at the top of the barrier. We assume that only one subband is occupied, so the directed moments can be evaluated to find

$$n_L^+(0) = \frac{N_{1D}}{2} F_{-1/2}(\eta_F)$$ (5.4a)

$$n_L^-(0) = \frac{N_{1D}}{2} F_{-1/2}(\eta_F - U_D)$$ (5.4b)

$$I^+ = \frac{q k_B T_L}{\pi \hbar} F_0(\eta_F)$$ (5.4c)

$$I^- = \frac{q k_B T_L}{\pi \hbar} F_0(\eta_F - U_D)$$ (5.4d)

$$\upsilon^+(0) = \upsilon_T \frac{F_0(\eta_F)}{F_{-1/2}(\eta_F)}$$ (5.4e)

$$\upsilon^-(0) = \upsilon_T \frac{F_0(\eta_F - U_D)}{F_{-1/2}(\eta_F - U_D)},$$ (5.4f)

where $\eta_F = (E_F - \varepsilon(0))/k_B T_L$, $U_D = q V_D / k_B T_L$, and $\upsilon_T = \sqrt{2 k_B T_L / \pi m^*}$ as before. The one-dimensional effective density of states is

$$N_{1D} = \sqrt{\frac{2 m^* k_B T_L}{\pi \hbar^2}}.$$ (5.4g)

It is important to note that the expressions for the directed currents are independent of the bandstructure, because when converting the sum over k-states to an integral over energy, the density of states cancels with the velocity in Eqns. (5.2c) and (5.2d).

5.2.1 Evaluation of the I-V characteristic

To evaluate the *I-V* characteristic of a nanowire transistor, we use the approach of Sec. 3.5. To keep things simple, we will neglect two-dimensional electrostatics. We begin by relating the potential at the top of the barrier to the gate voltage using an expression analogous to Eqn. (3.5),

$$\varepsilon_1(0) = -qV_G + \frac{q^2 n_L}{C_{ins}},$$

(5.5a)

which can be written as

$$\eta_F = \frac{(V_G - V_T)}{k_B T_L / q} - \frac{q^2 N_{1D}}{2k_B T_L C_{ins}} \{F_{-1/2}(\eta_F) + F_{-1/2}(\eta_F - U_D)\},$$

(5.5b)

where $V_T \equiv -E_F / q$. Equation (5.5b) is a nonlinear equation that can be solved for η_F at the given gate and drain voltages

To compute the *I-V* characteristic, we proceed as follows. First assume a source Fermi level, which is equivalent to setting a value of the off-current or threshold voltage. Next, solve Eqn. (5.5b) for η_F for a given drain and gate voltage. Then, compute the current from Eqns. (5.4c) and (5.4d) as

$$I_D = (I^+ - I^-)$$
$$= \frac{2qk_B T_L}{h} \{F_0(\eta_F) - F_0(\eta_F - U_D)\}.$$

(5.6)

Equations (5.5) and (5.6) can be readily solved to plot the *I-V* characteristics of a nanowire MOSFET. Before we do so, however, we can establish some of the important features of the *I-V* characteristic through simplified analytical solutions.

5.2.2 The I-V characteristics for nondegenerate carrier statistics

Traditional MOSFET analysis assumes nondegenerate carrier statistics (even though that assumption is questionable above threshold). Let's do the same for the nanowire MOSFET. Recall that for nondegenerate conditions, $F_j(\eta_F) \to e^\eta$ for any order, j, of the Fermi-Dirac integral. From Eqn. (5.6), we find the drain current as

$$I_D = \frac{2qk_BT_L}{h} e^{\eta_F} \left(1 - e^{-qV_{DS}/k_BT}\right). \tag{5.7}$$

To relate η_F to the gate voltage, we must solve Eqn. (5.5b), which under nondegenerate conditions reduces to

$$\eta_F = \frac{(V_{GS} - V_T)}{k_BT_L/q} - \frac{q^2 N_{1D}}{2C_{ins}} \frac{e^{\eta_F}}{k_BT_L} \left(1 + e^{-qV_{DS}/k_BT_L}\right). \tag{5.8}$$

Equation (5.8) has no analytical solution, but we can get an approximate solution under two conditions. Under subthreshold conditions, the charge in the channel is small, so Eqn. (5.8) becomes

$$\eta_F \approx \frac{(V_{GS} - V_T)}{k_BT_L/q} \tag{5.9}$$

which can be inserted into Eqn. (5.7) to find

$$I_D = \frac{2qk_BT_L}{h} e^{q(V_{GS} - V_T)/k_BT_L} \left(1 - e^{qV_{DS}/k_BT}\right), \tag{5.10}$$

which is just like the result for a traditional MOSFET [recall Eqn. (2.57)].

Above threshold, both terms on the right hand side of Eqn. (5.9) are large, so we can assume

$$0 \approx \frac{(V_{GS} - V_T)}{k_BT_L/q} - \frac{q^2 N_{1D}}{2C_{ins}} \frac{e^{\eta_F}}{k_BT_L} \{1 + e^{-qV_{DS}/k_BT_L}\}, \tag{5.11}$$

which can be solved for e^{η_F} and inserted into Eqn. (5.7) to find

$$I_D = C_{ins} \upsilon_T (V_{GS} - V_T) \frac{\left(1 - e^{qV_{DS}/k_BT}\right)}{\left(1 + e^{qV_{DS}/k_BT}\right)}, \tag{5.12}$$

where $\upsilon_T = \sqrt{2k_BT_L/\pi m^*}$ as before. Equation (5.12) is just like the corresponding result for a ballistic, planar MOSFET (recall eqn. 3.7b). There is an implicit assumption, in eqn. (5.12), however, that $C_Q \gg C_{ins}$, as it typically is for a MOSFET, but this condition may not always hold for a nanowire MOSFET.

5.2.3 The I-V characteristic for degenerate carrier statistics

For highly degenerate conditions, $\eta_F >> 1$ and $F_0(\eta_F) \to \eta_F$. We examine three device parameters – the channel conductance, the on-current, and the transconductance. By taking the degenerate limit in Eqn. (5.6), we find

$$I_D = \frac{2q^2}{h} V_D = G_Q V_D,$$ (5.13)

so the channel conductance is just the quantum conductance, as should have been expected. From the perspective of a traditional MOSFET, however, the result is unusual. The linear region current of a MOSFET is

$$I_D = \left(\frac{W}{L}\right) \mu_{eff} C_{ox} \left(V_G - V_T\right) V_D = g_d V_D,$$ (5.14)

so the channel conductance, g_d, of a MOSFET is proportional to $(V_G - V_T)$. In the quantum wire MOSFET, however, Eqn. (5.13) shows that the channel conductance is independent of the gate voltage (as long as the gate voltage is large enough to make the carriers degenerate), and small enough so that only one subband is populated

According to Eqn. (5.6), the drain current under high drain bias is

$$I_D = \frac{2q k_B T_L}{h} F_0(\eta_F).$$ (5.15)

Assuming high gate bias (or low temperatures), the Fermi-Dirac integral simplifies so

$$I_D(\text{on}) = \frac{2q^2}{h} \frac{\left(E_F - \varepsilon_1(0)\right)}{q}.$$ (5.16)

To relate the on-current to the gate voltage, we need to solve Eqn. (5.5), which means that we need to evaluate n_L for degenerate conditions. For degenerate conditions with a high drain bias, Eqn. (5.2a) becomes

$$n_L = n_L^+ = \frac{k_F}{\pi} = \frac{\sqrt{2m^* \left(E_F - \varepsilon_1(0)\right)}}{\pi \hbar},$$ (5.17)

which can be used in Eqn. (5.5), to find

$$\varepsilon_1(0) = -qV_G + \frac{q}{C_{ins}} \frac{\sqrt{2m^*\left(E_F - \varepsilon_1(0)\right)}}{\pi\hbar} \tag{5.18}$$

or

$$\frac{E_F - \varepsilon_1(0)}{q} = (V_G - V_T) - \frac{q^2}{C_{ins}} \frac{\sqrt{2m^*\left(E_F - \varepsilon_1(0)\right)}}{\pi\hbar}, \tag{5.19}$$

where $V_T \equiv -E_F/q$.

Equation (5.19) is a quadratic equation for $\sqrt{E_F - \varepsilon_1(0)}$, but there is another way to solve it. Recall that the quantum (or semiconductor) capacitance is

$$C_Q = \frac{\partial(q\,n_L)}{\partial\left(-\varepsilon_1(0)/q\right)} = \frac{q^2\sqrt{2m^*}}{\pi\hbar}\left(E_F - \varepsilon_1(0)\right)^{-1/2}, \tag{5.20}$$

which can be used to re-express Eqn. (5.19) as

$$\frac{E_F - \varepsilon_1(0)}{q} = \frac{V_G - V_T}{1 + C_Q/C_{ins}}. \tag{5.21}$$

Finally, according to Eqn. (5.16), the on-current is

$$I_D(on) = \frac{2q^2}{h} \frac{(V_G - V_T)}{1 + C_Q/C_{ins}}. \tag{5.22}$$

Since C_Q is gate voltage dependent, Eqn. (5.22) is not quite as simple as it looks.

We can write Eqn. (5.22) in a more familiar form as

$$I_D(on) = \frac{2q^2}{h} \frac{C_{GS}}{C_Q} (V_G - V_T), \tag{5.23a}$$

where $C_{GS} = C_{ins}C_Q/\left(C_{ins} + C_Q\right)$ is the gate to source capacitance. Using Eqn. (5.20), we can show that Eqn. (5.23a) is

$$I_D(on) = C_{GS}\upsilon^+(V_G - V_T), \tag{5.23b}$$

where

$$v^+ = \sqrt{\frac{(E_F - \varepsilon_1(0))}{2m^*}} \qquad (5.24)$$

is the average, x-directed velocity of the degenerate electron gas. Equation (5.23b) is the expected result for a MOSFET.

It is also interesting to examine Eqn. (5.22) in the quantum capacitance limit where $C_Q \ll C_{ins}$ and some unexpected things occur. In this case, the on-current becomes

$$I_D(\text{on}) = \frac{2q^2}{h}(V_G - V_T), \qquad (5.25)$$

which has a simple interpretation. Since there is no voltage drop across the insulator, the gate voltage directly controls the position of the Fermi level. At this limit, the transconductance is

$$g_m = \frac{2q^2}{h} = g_d \qquad (5.26)$$

In a traditional MOSFET, there is no simple relation between the transconductance and the channel conductance, but in a ballistic nanowire transistor at $T_L = 0K$, both are equal to the quantum conductance.

5.2.4 Numerical results

The discussion in the previous subsections establishes some general features of nanowire MOSFETs, and they demonstrate that such devices can behave differently from traditional MOSFETs. To obtain analytical expressions, we had to assume Boltzmann statistics or full degeneracy. To see what characteristics are expected at room temperature, we need to perform a numerical calculation as discussed in Sec. 5.2.1. For these calculations, we assume a $t_{wire} = 1$ nm silicon nanowire with and effective mass of $m^* = 0.19m_0$. For the insulator, we assume SiO$_2$ with $t_{ins} = 1$ nm. The coaxial gate geometry and ballistic transport will give us an upper limit assessment of the performance of this nanowire transistor. We assume a valley degeneracy of two and a drain capacitance that is 0.14 of the gate capacitance (2D electrostatics are included as discussed in Sec. 3.5.2). The threshold voltage, $V_T \equiv -E_F/q$, is 0.32V.

Figure 5.2 shows the computed $\log(I_D)$ vs. V_{GS} transfer characteristic and the I_D vs. V_{DS} common source characteristic at room temperature, and Fig. 5.3 is the same plot at $T_L = 77$K. At room temperature, the common source characteristic shows a channel conductance that increases with gate voltage, just as for a conventional MOSFET and not as predicted by Eqn. (5.13). The reason is that at room temperature and for moderate gate overdrive, the fully-degenerate assumption used to derive Eqn. (5.13) is not valid. At 77K, however, we see that the channel conductance is, indeed, $2q^2/h$ nearly independent of gate bias.

a)

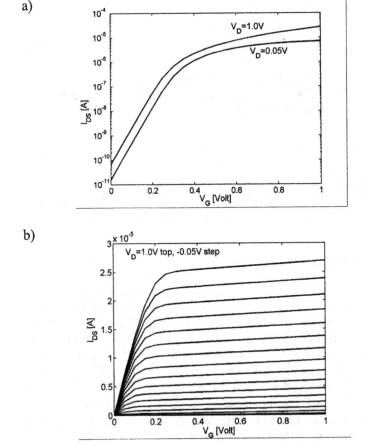

b)

Figure 5.2   The computed I-V characteristics of an idealized, ballistic nanowire MOSFET at $T_L = 300$K. (a) $\log(I_D)$ vs. V_{GS} and (b) I_D vs. V_{DS}.

a)

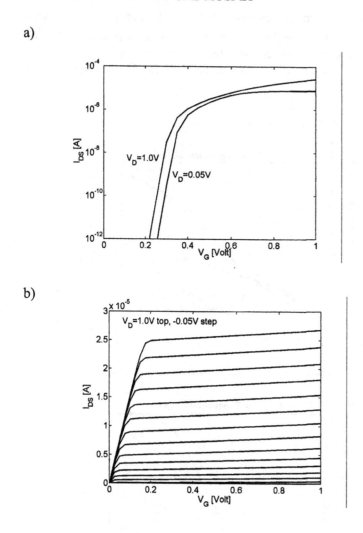

b)

Figure 5.3   The computed *I-V* characteristics of an idealized, ballistic nanowire
                  MOSFET at $T_L = 77\text{K}$.   (a) $\log(I_D)$ vs. V_{GS} and (b) I_D vs. V_{DS}.

Figure 5.4a is a plot of the low V_{DS} channel conductance and high V_{DS}
transconductance vs. gate voltage at room temperature and Fig. 5.4b is the
same plot at 77K. We see that for high gate voltages, the channel

conductance does, indeed, approach $2q^2/h$. If higher subbands were
included, we would see the channel conductance increase in steps of $2q^2/h$.
Figure 5.4 also shows, however, that the transconductance is smaller than
the channel conductance. According to Eqn. (5.22), this implies that
$C_Q > C_{ins}$.

a)

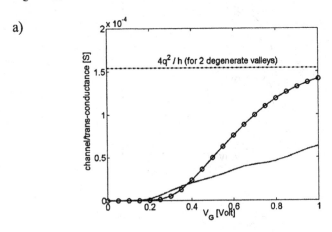

b)

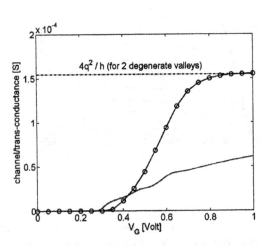

Figure 5.4b   (a) The low-V_{DS} channel conductance (circles) and the high-V_{DS}
                     transconductance (line) of an idealized, ballistic nanowire MOSFET at T_L
                     = 300K. (b) The low-V_{DS} channel conductance (circles) and the high-V_{DS}
                     transconductance (line) of an idealized, ballistic nanowire MOSFET at T_L
                     = 77K.

Figure 5.5 is a plot of C_Q vs. V_{GS} for high drain bias with the constant value of C_{ins} also shown. Both room temperature and $T_L = 77K$ results are shown. The high drain bias quantum (or semiconductor) capacitance is evaluated from

$$C_Q = \frac{\partial(qn_L^+)}{\partial(-\varepsilon(0)/q)} = \frac{1}{k_B T_L}\frac{\partial(qn_L^+)}{\partial \eta_F} = \frac{N_{1D}}{2k_B T_L}F_{-3/2}(\eta_F), \qquad (5.26)$$

where Eqn. (5.4a) was used. Figure 5.5 shows that this nanowire transistor does not operate at the quantum capacitance limit where $g_m = g_d$ because $C_Q > C_{ins}$ over the bias range of interest. Recall from Eqn. (2.23) that C_Q is proportional to the density-of-states at the Fermi level averaged over a $k_B T_L$ to account for thermal broadening. At the band edge, the 1D density-of-state goes to infinity, which explains the rise in C_Q near the band edge, as observed in Fig. 5.5. At low temperatures, the rise is more dramatic because the thermal broadening is reduced.

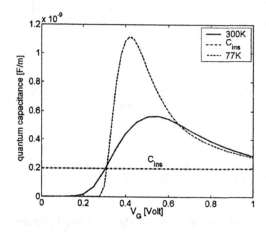

Figure 5.5   The semiconductor (or quantum) capacitance vs. gate voltage for an idealized, ballistic nanowire MOSFET at $T_L = 300K$ (solid line) and $T_L = 77K$ (dashed line).

It is also of interest to examine the carrier density at the top of the barrier to see how it varies with drain bias. As discussed in Chapter 3, for a traditional MOSFET, which operates well below the quantum capacitance limit, $n_L(0) = n_L^+(0) + n_L^-(0)$ is controlled by the gate voltage and is independent of the drain bias. At the quantum capacitance limit, however, the increasing drain bias empties the negative k-states and $n_L^+(0)$ is fixed, so $n_L(0)$ at high drain bias is one-half of its value at low drain bias. Figure 5.6, a plot of $n_L(0)$ vs. V_{DS} at room temperature, shows that this nanowire MOSFET operates closer to the traditional MOS limit, $C_Q \gg C_{ins}$ than to the quantum capacitance limit where $C_Q \ll C_{ins}$.

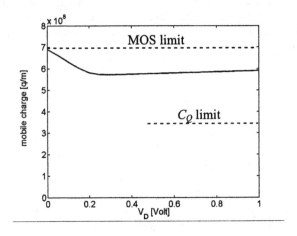

Figure 5.6   The carrier density at the top of the barrier vs. V_{DS} for an idealized, ballistic nanowire MOSFET at $T_L = 300$K.

Finally, it is of interest to examine the high V_{DS} injection velocity vs. gate voltage. Figure 5.7 shows that carrier degeneracy make the injection velocity gate-voltage dependent and that high values are possible.

Having seen the expected characteristics of a silicon-like ballistic nanowire transistor, we are now ready to examine a MOSFET-like carbon nanotube transistor.

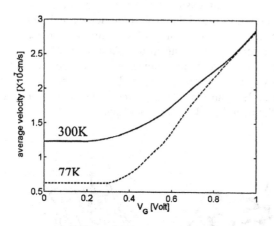

Figure 5.7 The high-V_{DS} channel injection velocity vs. V_{GS} of an idealized, ballistic nanowire MOSFET at $T_L = 300$K and at $T_L = 77$K.

5.3 Carbon Nanotubes

Since their discovery in the early 1990's [5.6, 5.7], single-walled carbon nanotubes (SWNT's) have been the subject of intense research. They display remarkable electronic, mechanical, and thermal properties [5.8] and are, therefore, promising for new device applications. Carbon nanotube (CNT) electronics is one such promising application [5.9]. Although the field is still new, field-effect transistors [5.1, 5.2, 5.3] and circuits [5.10, 5.11, 5.12], pn junctions [5.12], bio-sensors [5.13], and optical emitters [5.14] have all been demonstrated. Carbon nanotube electronics is both scientifically and technologically interesting.

This chapter is an introduction to carbon nanotube electronics; it is not a comprehensive look at the field. We will examine carbon nanotube field-effect transistors (CNTFETs) and use them as a specific example to see how the ideas developed in the previous section are applied to much different material systems. We begin by reviewing the basic electronic properties of CNTs, which include band structures and the density-of-states (DOS).

The bandstructure of a SWNT can be understood in terms of the bandstructure of graphene, a two-dimensional sheet of carbon atoms that is rolled up to form a nanotube. Graphite consists of layers of carbon atoms arranged in a hexagonal lattice like chicken wire. Each layer, known as graphene, consists of a two-dimensional sheet of carbon atoms in the honeycomb structure shown in Fig. 5.8a. The three nearest neighbor bonds for each carbon atom are made up of the carbon s orbital and two of the three p orbitals. The strong sp^2 bonding in the plane is what leads to the superior mechanical properties and electromigration resistance of SWNTs. Although the bonding within a layer is strong, the layers themselves are held together by weak (so-called van der Waals) forces. The easy removal of layers is what makes the pencil mark on a piece of paper.

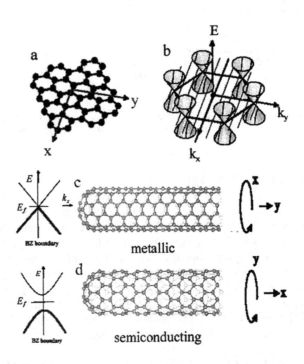

Figure 5.8   Carbon nanotubes can be viewed as a rolled-up graphene sheet. The periodic boundary condition only allows quantized wave vectors around the circumferential direction, which generates one-dimensional bands for carbon nanotubes. (Reproduced with permission from [5.9])

Graphene has an unusual bandstructure; the conduction and valence bands meet at the corners of the Brillouin zone as shown in Fig. 5.8b (recall that the Brillouin zone of a hexagonal real space lattice is also hexagonal). If $E = 0$ is the energy at the intersection point, then at $T_L = 0K$ all states below $E = 0$ are filled and those above are empty. Because there are empty states at the Fermi level, the behavior is metallic. Graphene is known as a zero-gap semiconductor.

The simplest way to think of a SWNT is as a rolled-up sheet of graphene. Note that in this rolled-up geometry, all sp^2 carbon bonds are still satisfied, so there are no dangling bonds in the nanotube – in contrast to the situation for a typical semiconductor-insulator interface. One expects, therefore, that semiconductor-insulator interfaces should display low densities of interface states.

For the nanotube, we enforce a boundary condition around the circumference of the nanotube,

$$\vec{k} \bullet \vec{C} = 2\pi v, \tag{5.27}$$

where $\vec{C}$ is a vector in the x-y plane of the graphene sheet that becomes the circumference of the nanotube, and v is an integer. The allowed values of $\vec{k}$ are quantized by the boundary condition. Since there is no boundary condition to apply along the length of an infinitely long nanotube, the result is that the allowed k's are a series of lines in k-space as shown in Fig. 5.8b. Along the axis of the nanotube, any value of k is permitted. The spacing of the lines of allowed circumferential k is determined by v and by how the graphene sheet was rolled up as specified by $\vec{C}$. Two possibilities exist; either the allowed k-states do or do not pass through one of the Fermi points where the conduction and valence bands meet. If the first case occurs, we have a metallic nanotube (Fig. 5.8c), and if the second case occurs, a semiconducting nanotube (Fig. 5.8d).

5.4 Bandstructure of Carbon Nanotubes

5.4.1 Bandstructure of graphene

A simple way to calculate the E vs. k relation of a carbon nanotube, is to quantize the two-dimensional E vs. k of the graphene sheet along the circumferential direction of the nanotube. The first step, therefore, is to understand the band structure of a graphene sheet.

The graphene lattice is shown in Fig. 5.9a. Note that each carbon atom does not see the same environment, but each unit cell of two carbon atoms does. The two-dimensional graphene lattice can be created by translating a unit cell by the vectors $\vec{T} = n\vec{a}_1 + m\vec{a}_2$ with integer combinations of (n, m). Here $\vec{a}_1$ and $\vec{a}_2$ are the basis vectors of the lattice (also shown in Fig. 5.9),

$$\vec{a}_1 = \sqrt{3}a_{CC}\left(\frac{\sqrt{3}}{2}\hat{x} + \frac{1}{2}\hat{y}\right) = a_0\left(\frac{\sqrt{3}}{2}\hat{x} + \frac{1}{2}\hat{y}\right) \tag{5.28a}$$

$$\vec{a}_1 = \sqrt{3}a_{CC}\left(\frac{\sqrt{3}}{2}\hat{x} - \frac{1}{2}\hat{y}\right) = a_0\left(\frac{\sqrt{3}}{2}\hat{x} - \frac{1}{2}\hat{y}\right), \tag{5.28b}$$

where $a_0 = \sqrt{3}a_{cc}$ is the length of the basis vector, and $a_{CC} \approx 1.42 \overset{o}{A}$ is the nearest neighbor C-C bonding distance.

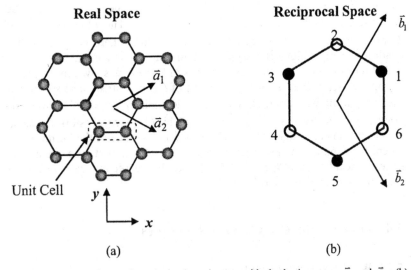

(a)   (b)

Figure 5.9   (a) The graphene lattice in real space with the basis vectors $\vec{a}_1$ and $\vec{a}_2$. (b) The first Brillouin zone of the reciprocal lattice with the basis vectors $\vec{b}_1$ and $\vec{b}_2$.

The 2s, $2p_x$, and $2p_y$ orbitals give rise to the so-called σ bonds of graphene. The σ bonds are only weakly coupled to the $2p_z$ orbitals, which produce the so-called π bonds, which give rise to the electronic properties of graphene. The $E(k)$ relation for graphene can be computed from a tight-binding model that uses one $2p_z$ orbital with nearest neighbor interactions. More detailed calculations including multiple orbitals and more neighboring atoms show that the one-orbital tight-binding approximation works well at the energy range near the Fermi point of the graphene sheet, which is the region of interest for electronic transport. The result of the tight-binding calculation is [5.8, 5.15, 5.16]

$$E(\bar{k}) = \pm t \sqrt{1 + 4\cos\left(k_x \frac{3a_{CC}}{2}\right)\cos\left(k_y \frac{\sqrt{3}a_{CC}}{2}\right) + 4\cos^2\left(k_y \frac{\sqrt{3}a_{CC}}{2}\right)} ,(5.29)$$

where $t \approx 3.0$ eV is the C-C bonding energy. The positive sign gives the conduction band and the negative sign the valence band. In contrast to silicon, the conduction and valence bands of the graphene sheet are mirror images of each other, which is advantageous for complementary devices. The plus sign gives rise to the so-called π^* anti-bonding levels and the minus sign to the π bonding levels. The two bands intersect at $E = 0$, and as indicated in Fig. 5.8b, this occurs at specific locations within the Brillouin zone.

The energy levels near the Fermi level are most important for electronic transport, so we need to know where these so-called Fermi points lie and how many of them there are. The basis vectors for the reciprocal lattice $\bar{b}_j$, as shown in Fig. 5.9 (b), satisfy

$$\bar{a}_i \cdot \bar{b}_j = 2\pi\delta_{ij}, \tag{5.30}$$

so using the real-space basis vectors from Eqns. (5.28), we find

$$\bar{b}_1 = \frac{4\pi}{3a_{CC}}\left(\frac{1}{2}\hat{x} + \frac{\sqrt{3}}{2}\hat{y}\right) = b_0\left(\frac{1}{2}\hat{x} + \frac{\sqrt{3}}{2}\hat{y}\right) \tag{5.31a}$$

$$\bar{b}_2 = \frac{4\pi}{3a_{CC}}\left(\frac{1}{2}\hat{x} - \frac{\sqrt{3}}{2}\hat{y}\right) = b_0\left(\frac{1}{2}\hat{x} - \frac{\sqrt{3}}{2}\hat{y}\right), \tag{5.31b}$$

where $b_0 = 4\pi/\sqrt{3}a_0$ is the length of the basis vector in reciprocal space. The wave vectors at the six corners of the Brillouin zone can be expressed in terms of b_1 and b_2 as

$$\bar{k}_{F1} = \left(\frac{2\pi}{3a_{CC}}, \frac{2\pi}{3\sqrt{3}a_{CC}} \right), \tag{5.32a}$$

$$\bar{k}_{F2} = \left(0, \frac{4\pi}{3\sqrt{3}a_{CC}} \right) \tag{5.32b}$$

$$\bar{k}_{F3} = \left(\frac{-2\pi}{3a_{CC}}, \frac{2\pi}{3\sqrt{3}a_{CC}} \right) \tag{5.32c}$$

$$\bar{k}_{F4} = \left(-\frac{2\pi}{3a_{CC}}, -\frac{2\pi}{3\sqrt{3}a_{CC}} \right) \tag{5.32d}$$

$$\bar{k}_{F5} = \left(0, -\frac{4\pi}{3\sqrt{3}a_{CC}} \right) \tag{5.32e}$$

$$\bar{k}_{F6} = \left(\frac{2\pi}{3a_{CC}}, -\frac{2\pi}{3\sqrt{3}a_{CC}} \right) \tag{5.32e}$$

where the numbering is shown in Fig. 5.9. By substituting $\bar{k}_F$ from Eqn. (5.32) into Eqn. (5.29), we can show that the energy at each of the six Fermi points is zero. There are two groups of Fermi points (the filled and open circles in Fig. 5.9). Within each group, all the points differ by a reciprocal lattice vector, so they are equivalent. The result is that there are only two independent Fermi points, which we can take to be

$$\bar{k}_F = \left(0, \pm \frac{4\pi}{3\sqrt{3}a_{CC}} \right). \tag{5.32f}$$

When we derive the $E(k)$ relation about a Fermi point, we will have to remember that there is a valley degeneracy of two.

5.4.2 Physical structure of nanotubes

The formation of a carbon nanotube can be viewed as the rolling up of a graphene sheet. As shown in Fig. 5.10, the chiral vector specifies the direction of the roll-up,

$$\vec{C} = n\vec{a}_1 + m\vec{a}_2 \equiv (n,m). \tag{5.33}$$

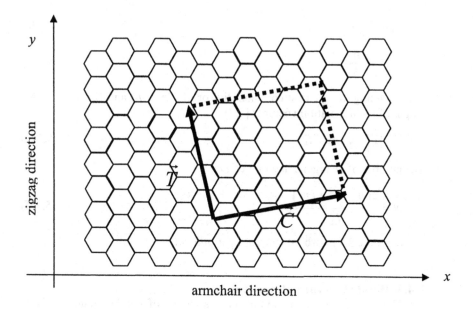

y

zigzag direction

armchair direction

x

Figure 5.10   Illustration of how a graphene sheet is rolled up into a nanotube. The chiral
vector, $\vec{C}$, specifies the circumference of the nanotube and $\vec{T}$ is a vector
along the axis of the nanotube. The length of $\vec{T}$ defines the unit cell for the
nanotube (the region within the rectangle). If $\vec{C}$ lies along the x-axis, we have
an armchair nanotube and if $\vec{C}$ lies along the y-axis, a zigzag nanotube,.

Once the chiral vector is specified, several parameter of the nanotube
follow [5.8]. From the length of the chiral vector, we find the diameter of
the nanotube as [5.8]

$$d = \frac{\sqrt{3}a_{CC}}{\pi}\sqrt{n^2 + m^2 + nm} \ .$$
(5.34a)

The translation vector for the nanotube is [5.8]

$$\vec{T} = t_1\,\vec{a}_1 + t_2\,\vec{a}_2 ,$$
(5.34b)

where

$$t_1 = \frac{2m+n}{d_R} \tag{5.34c}$$

$$t_2 = -\frac{2n+m}{d_R}, \tag{5.34d}$$

with d_R being the greatest common divisor of $(2m + n)$ and $(2n + m)$. The length of the translation vector is [5.8]

$$T = \left| \vec{T} \right| = \sqrt{3}\pi d / d_R. \tag{5.34e}$$

The number of hexagons in a unit cell is

$$N = \frac{2\left(m^2 + n^2 + nm\right)}{d_R}, \tag{5.34f}$$

and there are $2N\,p_z$ orbital electrons per unit cell.

5.4.3 Band structure of nanotubes

The simplest way to obtain the band structure of a carbon nanotube is to begin with the bandstructure of graphene, Eqn. (5.29) and apply periodic boundary conditions along the circumference of the nanotube. From Eqns. (5.27) and (5.33), we get

$$k_x(n+m)\frac{3a_{CC}}{2} + k_y(n-m)\frac{\sqrt{3}a_{CC}}{2} = 2\pi v. \tag{5.35}$$

Equation (5.35) is an equation for a set of parallel lines. If one of these lines passes through a Fermi point, we have a metallic nanotube, if not, a semiconducting nanotube. In order to pass through the Fermi point at $(0, 4\pi/3\sqrt{3}\,a_{CC})$, Eqn. (5.35) states that

$$\frac{(n-m)}{3} = v \tag{5.36}$$

so when $(n - m)$ is a multiple of 3, one P_z orbital model says that the nanotube is metallic.

Let's be specific by examining *armchair nanotubes*, which are specified by (n, n) – a roll-up direction along the x-axis. Referring to Fig. 5.10, we see that the carbon atoms lie in an "armchair" pattern along the x-direction. Equations (5.33), (5.34b), and (5.34f) give

$$\vec{C} = (n,n) = n3a_{CC}\,\hat{x}.\tag{5.37a}$$

$$\vec{T} = \vec{a}_1 - \vec{a}_2 = \sqrt{3}a_{CC}\hat{y}\tag{5.37b}$$

and

$$N = 2n.\tag{5.37c}$$

Because there are $2n$ hexagons in the unit cell of an armchair nanotube, we get $2n$ pairs of energy bands with allowed wavectors in the x-direction given from Eqn. (5.27) as

$$k_x^v = \frac{2\pi v}{n3a_{CC}}\qquad (v = 1, 2, ..., 2n).\tag{5.38}$$

As shown in Fig. 5.11. Equation (5.38) defines a series of parallel lines of constant k_x. It is clear that for the subband $v = 0$, the allowed k's will pass through the Fermi points, $\left(0, \pm 4\pi/3\sqrt{3}a_{CC}\right)$, regardless of the value of n, so all armchair nanotubes are metallic. When Eqn. (5.38) is inserted into the graphene bandstructure, Eqn. (5.29), we find the $E(k)$ relation for an armchair nanotube as

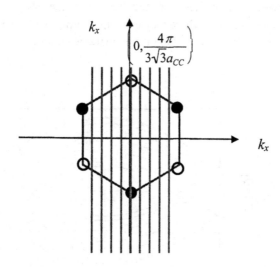

lines of allowed wavevectors

Figure 5.11   Illustration of the lines of allowed k's for an armchair nanotube.

$$E(k_t) = \pm t \sqrt{1 + 4\cos\left(\frac{v\pi}{n}\right)\cos\left(\frac{k_t\sqrt{3}a_{CC}}{2}\right) + 4\cos^2\left(\frac{k_t\sqrt{3}a_{CC}}{2}\right)}. \qquad (5.39)$$

In Eqn. (5.39), k_t refers to the magnitude of k along the axis of the nanotube (k_y for an armchair nanotube) and the range of k_t is $\left(-\pi/T < k_t < \pi/T\right) = \left(-\pi/\sqrt{3}a_{CC} < k_t < \pi/\sqrt{3}a_{CC}\right)$.

Figure 5.12 is a plot of $E(k_t)$ for a (7,7) armchair nanotube. The location of the Fermi level is $E = 0$, and there are 14 ($N = 2n$) conductions bands and 14 valence bands (some are degenerate, so the number looks smaller). A valence and conduction band cross at the Fermi point, $k_t = 2\pi/3\sqrt{3}a_{CC}$, which is 2/3 of the distance from the zone center to the zone boundary. They cross at the Fermi level, and the energy bands are symmetric for $\pm k_t$. Armchair nanotubes exhibits metallic conduction because an infinitesimal excitation puts carriers in the conduction band.

More generally, all (n, n) armchair nanotubes produce $2n$ conduction and $2n$ valence bands. All armchair nanotubes have a band degeneracy between the highest valence band and the lowest conduction band at $k_t = 2\pi/3\sqrt{3}a_{CC}$ where they cross the Fermi level. All armchair nanotubes, therefore, are expected to display metallic conduction.

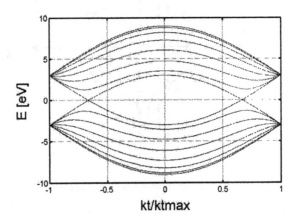

Figure 5.12 $E(k)$ relation for a (7, 7) armchair nanotube.

Next, we consider *zigzag nanotubes*, which can be described by $(n, -n)$, which, given the hexagonal symmetry is equivalent to $(n, 0)$. For zigzag nanotubes, Eqns. (5.33), (5.34b), and (5.34f) give

$$\vec{C} = (n, -n) = n\sqrt{3}a_{CC}\,\hat{y}. \qquad (5.40a)$$

$$\vec{T} = 3a_{CC}\,\hat{x} \qquad (5.40b)$$

and

$$N = 2n. \qquad (5.40c)$$

Again, we get $2n$ pairs of energy bands this time with allowed wavevectors in the y-direction given from Eqn. (5.27) as

$$k_y^\nu = \frac{2\pi\nu}{n\sqrt{3}a_{CC}} \qquad (\nu = 1, 2, ..., 2n). \qquad (5.41)$$

As shown in Fig. 5.13, Eqn. (5.13) defines a series of parallel lines of constant k_y. One of the lines of allowed k_y will pass through the Fermi points, $\left(0, \pm 4\pi/3\sqrt{3}a_{CC}\right)$, only if n is a multiple of 3. When Eqn. (5.41) is inserted into the graphene bandstructure, Eqn. (5.29), we find the $E(k)$ relation for an $(n, 0)$ zigzag nanotube as

$$E(k_t) = \pm t\sqrt{1 + 4\cos\left(\frac{3k_t a_{CC}}{2}\right)\cos\left(\frac{\nu\pi}{n}\right) + 4\cos^2\left(\frac{\nu\pi}{n}\right)}. \qquad (5.42)$$

In Eqn. (5.42), k_t refers to the magnitude of k along the axis of the nanotube (k_y for this zigzag nanotube) and the range of k_t is $\left(-\pi/T < k_t < \pi/T\right) = \left(-\pi/3a_{CC} < k_t < \pi/3a_{CC}\right)$.

Figure 5.14a is a plot of $E(k)$ for a (12,0) zigzag nanotube and Fig. 5.14b for a (13,0) zigzag nanotube. Again, the location of the Fermi level is $E = 0$, and there are $N = 2n$ pairs of conduction and valence band. For the (12,0) nanotube, the highest valence band and the lowest conduction band cross at $k = 0$. For an $(n, 0)$ zigzag nanotube, the energy gap is zero at $k = 0$ when n is a multiple of 3, and such nanotubes are metallic. Figure 5.7 shows that when n is not a multiple of 3, then a bandgap occurs at $k = 0$, and such zigzag nanotubes are semiconducting. In addition, when n is even, dispersionless bands occur (i.e. bands for which $dE/dk = 0$).

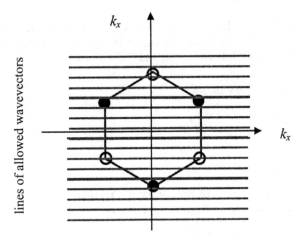

Figure 5.13   Illustration of the lines of allowed k's for a zigzag nanotube.

For a general roll-up vector, $\vec{C}$, we have a *chiral* nanotube. In this case, the lines of allowed k's run at an angle in the $k_x - k_y$ plane. When $(n - m)$ is a multiple of 3, then one of them intersects a Fermi point, and the nanotube is metallic [5.8]. By considering all of the possible ways that a nanotube can be rolled up, one can show that about 1/3 of the possible nanotubes are metallic and the rest semiconducting [5.8]. We saw that for metallic zigzag and armchair nanotubes, the band crossings occurred at $k_t = 0$ or at $k_t = \pm 2\pi/3T$ respectively. The same two possibilities exist in the general chiral case. Metallic nanotubes with band crossings at $k_t = 0$ are known as metal-1 nanotubes and those with band crossings at $k_t = \pm 2\pi/3T$ are known as metal

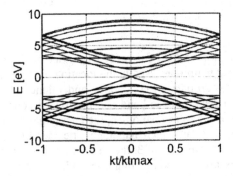

Figure 5.14a $E(k)$ relation for a (12, 0) zigzag nanotube.

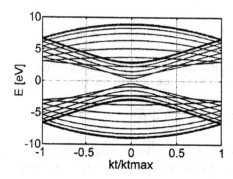

Figure 5.14b $E(k)$ relation for a (13, 0) zigzag nanotube.

5.4.4 Band structure near the Fermi points

Equation (5.29) gives an analytical expression for the $E(\vec{k})$ relation throughout the entire Brillouin zone of graphene. Since we are interested in $E(\vec{k})$ near a Fermi point, we simplify Eqn. (5.29) by using a Taylor series expansion for the cosine function near the Fermi point to find

$$E(\vec{k}) = \frac{3a_{cc}t}{2}\left|\vec{k}-\vec{k}_F\right| = \frac{3a_{cc}t}{2}\sqrt{(k_x-k_{Fx})^2+(k_y-k_{Fy})^2} \,, \qquad (5.43)$$

which shows $E(\vec{k}-\vec{k}_F)$ is linear and isotropic near a Fermi point (as shown in Fig. 5.8b). This linear approximation agrees with Eqn. (5.29) within an energy range ~1 eV near the Fermi point. Due to its mathematical simplicity, Eqn. (5.43) is useful for deriving analytical forms of electronic properties such as the density-of-states and bandgap of carbon nanotubes [5.17].

For nanotubes, we are also most interested in $E(k)$ near the Fermi-points. If the origin of the reciprocal lattice is translated to the Fermi point, the wave vector in the new coordinate system becomes

$$\vec{k}' = \vec{k}-\vec{k}_F = k_c'\,\hat{a}+k_t'\,\hat{t} \,, \qquad (5.44)$$

where k_c' is the component along the circumferential direction, which is quantized by the periodic boundary condition,

$$k'_{c,v} = (\vec{k} - \vec{k}_F) \cdot \hat{\partial} = \frac{\vec{k} \cdot \vec{C} - \vec{k}_F \cdot \vec{C}}{\left|\vec{C}\right|} = \frac{2}{3d}[3v - (n-m)], \qquad (5.45)$$

and k'_t is the wavevector along the axis of the nanotube.

Based on Eq. (5.43), the approximation for the graphene bandstructure near the Fermi point, the $E(k)$ relation of the CNT is

$$E(\vec{k}) = \frac{3a_{cc}\,t}{2}\left|\vec{k}'\right| = \frac{3a_{cc}\,t}{2}\sqrt{k'^2_{c,v} + k'^2_t} \qquad (5.46)$$

The lowest band of the CNT is determined by the minimum value of $k_{c,v}$. There are two cases to consider; when $(n - m)$ is a multiple of 3, then the nanotube is metallic, otherwise it is semiconducting.

Consider a metallic nanotube first. The minimum circumferential wave vector is $k'_{c,v} = 0$ which occurs when $v = (n-m)/3$. From Eqn. (5.46), then $E(k)$ relation for a metallic nanotube becomes

$$E = \pm\frac{3a_{cc}\,t}{2}k'_t, \qquad (5.47)$$

which is a one-dimensional linear dispersion relation independent of (n, m) as shown in Fig. 5.15

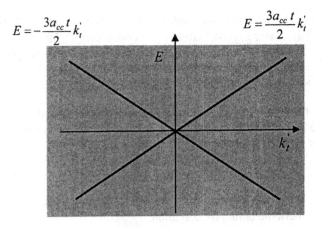

Figure 5.15   The $E(k)$ relation near the Fermi point for a metallic CNT.

The 1D density of states for this metallic CNT is

$$D(E) = D_0 = \frac{8}{3\pi a_{cc} t},$$ (5.48)

which is independent of energy.

Consider next a semiconducting nanotube, which occurs when $(n - m)$ is not a multiple of 3. In this case, the minimum magnitude of the circumferential wavevector is found from Eqn. (5.45) as

$$k_{c,v} = \frac{2}{3d}$$ (5.49)

(recall that d is the diameter of the CNT). By substituting Eqn. (5.49) into the linear $E(k)$ approximation for the CNT Eqn. (5.41), we get

$$E(k_t') = \pm \frac{3 a_{cc} t}{2} \sqrt{k_t'^2 + (2/3d)^2},$$ (5.50)

which is a one-dimensional $E(k)$ for the semiconducting nanotube. Note that the dispersion relation is independent of n and m (just as for the metallic nanotube). Near the band minimum, all metal nanotubes are alike and all semiconducting nanotubes are alike. Figure 5.16 is a plot of $E(k)$ for a semiconducting nanotube. Note that $E(k)$ is non parabolic.

According to Eqn. (5.50), the conduction and valence bands of a semiconducting CNT are mirror images of each other and the bandgap is

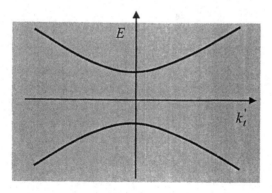

Figure 5.16   The $E(k)$ relation near the Fermi point for a semiconducting CNT.

$$E_G = \frac{2a_{cc}\,t}{d} \approx \frac{0.8}{d(\text{in nm})} \text{ eV}. \tag{5.51a}$$

Based on this simple derivation, the $E(k)$ relation and the bandgap are functions of the CNT diameter alone and do not depend on the specific values of n and m. In addition to the first subband with a bandgap given by Eqn. (5.51a), there are higher order subbands that may be populated as well. The bandgap of the ith subband is

$$E_{Gi} = \frac{2a_{cc}\,t}{d} \times \left(\frac{6i - 3 - (-1)^i}{4} \right) \qquad i = 1, 2, 3, \dots \tag{5.51b}$$

From Eqn. (5.50), we find the one-dimensional conduction band density of states for one semiconducting band as

$$D(E) = D_0 \frac{|E|}{\sqrt{E^2 - (E_G/2)^2}} \Theta(|E| - E_G/2), \tag{5.52}$$

where D_0 is the constant metallic band density of states and $\Theta(x)$ is the step function which equals 1 for $x > 0$ and 0 otherwise. Figure 5.17 plots the density of states for a (12, 0) and a (13, 0) nanotube. Near the Fermi level ($E = 0$), the (12, 0) nanotube shows the constant density of states expected from Eqn. (5.48). The density of states near the Fermi level is zero for the (13, 0) nanotube because of the presence of a bandgap. For both cases, we see contributions to the DOS from higher subbands that arise from the different values of v.

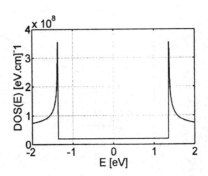

Figure 5.17a   The density of states vs. energy near the Fermi point for a (12, 0) zigzag metallic CNT.

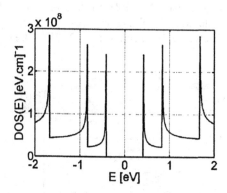

Figure 5.17b   The density of states vs. energy near the Fermi point for a (13, 0) zigzag
semiconducting CNT.

5.5 Carbon Nanotube FETs

Single walled carbon nanotubes were discovered in 1993 [5.6], and only
a few years later carbon nanotube field-effect transistors (CNTFETs) [5.1,
5.2, 5.3] and circuits [5.10, 5.11] were realized. The science and technology
of CNTFETs is still at an early stage, but they do show promise. They also
provide an opportunity to apply the theory of ballistic transistors to an
interesting new materials system. At this stage, it is hard to predict the
future of carbon nanotube electronics, but working on CNTFETs provides a
concrete context in which to learn how to treat electronic devices at the
molecular scale. The issues relating to contact, interfaces, transport, etc.
that must be addressed for CNTFETs are likely to be important for other
types of devices as well.

Before we discuss CNTFETs, it is important to realize that there are two
different ways to make a transistor. As shown in Fig. 5.18a, the first type is
a MOSFET-like device for which any charge demanded by the gate is
supplied from the contacts. We refer to such a device as a charge-
modulation transistor. The second type of transistor is shown in Fig. 5.18b.
In this case, there is a barrier at the contact, and the transistor operates by
modulating the width of the tunnel barrier and, therefore, the current that
flows in from the contact. We refer to this type of device as a transmission-
modulation transistor. It is also known as a Schottky barrier transistor and,
although it is an old idea [5.18], it continues to be explored [5.19]. One can,
therefore, conceive of two types of CNTFETs – charge modulation

(MOSFET-like) CNTFETs and transmission-modulation (Schottky barrier) CNTFETs. Although most CNTFETs to date behave as Schottky barrier CNTFETs [5.20], the MOSFET-like device have recently been reported [5.21]. We begin, therefore, by discussing MOSFET-like CNTFETs.

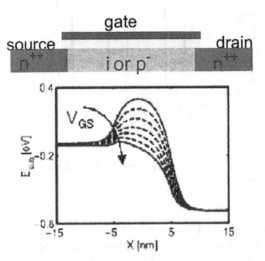

Figure 5.18a   Illustration of a MOSFET-like transistor. The gate voltage pushes down the source to channel barrier and allows any amount of charge demanded by the gate to enter the channel.

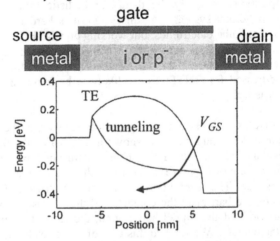

Figure 5.18b   Illustration of a Schottky barrier transistor. The gate voltage squeezes a barrier between the source and channel barrier, which increases the tunneling current through the barrier.

5.6 Carbon Nanotube MOSFETs

A MOSFET-like CNTFET can be described by the same theory presented in Sec. 5.3 for semiconductor nanowires MOSFETs. The current is independent of bandstructure, so Eqn. (5.6) still applies. Equation (5.5) still describes the electrostatics, but to relate n_L^+ and n_L^- to the Fermi levels, we need to evaluate the sums, Eqns. (5.2a) and (5.2b) with the bandstructure of a carbon nanotube as given by Eqn. (5.50) and (5.51),

$$E(k) = \frac{E_G}{2}\sqrt{1+\left(\frac{3kd}{2}\right)^2}. \tag{5.53}$$

Near $k = 0$, we can expand the $E(k)$ relation for small argument to find

$$E(k) \approx \frac{E_G}{2} + \frac{\hbar^2 k^2}{2m^*}, \tag{5.54a}$$

where

$$m^* = \frac{4\hbar^2}{9a_{CC}\,dt} \tag{5.54b}$$

or

$$\frac{m^*}{m_0} = \frac{0.08}{d(\text{nm})}. \tag{5.54c}$$

Unfortunately, the parabolic band assumption is not a good one for carbon nanotubes. Figure 5.19 compares the nanotube $E(k)$ relation from Eqn. (5.50) with a parabolic band assumption using the effective mass from Eqn. (5.54) with $d = 1$ nm. The parabolic band approximation is valid only very near the energy minimum, but away from the minimum, $E(k)$ is linear as in a metallic nanotube. The result is that analytical expression for n_L has to be replaced by numerical integrations.

The I-V characteristics of the MOSFET-like CNTFET are described by Eqns. (5.5) and (5.6), which describe the electrostatics and transport. To evaluate the carrier density, we evaluate Eqn. (5.2a)

$$n_L^+ = \frac{1}{L}\sum_{k>0} f_0(E_F) = \int_0^{E_{Top}} f_0(E_F) D_{CNT}(E)\,dE \tag{5.55}$$

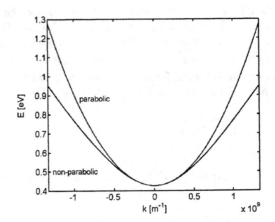

Figure 5.19   Comparison of $E(k)$ for a 1nm diameter nanotube as computed from Eqn. (5.50) and Eqn. (5.54).

where Eqn. (5.52) is used for the carbon nanotube density-of-states. Unfortunately, Eqn. (5.55) cannot be done analytically, or even in terms of tabulated functions (as for the Fermi-Dirac integral of order $-1/2$ when parabolic bands are assumed). For a given nanotube, however (i.e. for a given E_G and D_0),

$$n_L^+ = D_0 \int_0^\infty \frac{E}{\sqrt{E^2 - (E_G/2)^2}} \left(\frac{1}{1+e^{(E-E_F)/k_BT_L}} \right) dE$$

$$= \frac{N_{CN}}{2} \int_0^\infty \frac{(\xi + \xi_G/2)}{\sqrt{\xi^2 + \xi\xi_G}} \left(\frac{1}{1+e^{\xi-\eta_F}} \right) d\xi \qquad (5.56a)$$

where $N_{CN} = k_BT_LD_0$ and $\xi_G = E_G/k_BT_L$. The integral depends only on $\eta_F = (E_F - E_C)/k_BT_L$, so we can write the result as

$$n_L^+ = N_{CN}F_{CNT}(\eta_F), \qquad (5.56b)$$

where $F_{CNT}(\eta_F)$ simply stands for the value of the numerical integral. Similar considerations apply to n_L^- with η_F replaced by $\eta_F - U_D$. The result is that Eqn. (5.5b) becomes

$$\eta_F = \frac{(V_G - V_T)}{k_B T_L / q} + \frac{q^2 N_{CN}}{2k_B T_L C_{ins}} \{F_{CN}(\eta_F) + F_{CN}(\eta_F - U_D)\}.$$ (5.57a)

This equation must be solved iteratively, and the F_{CN} functions must be evaluated by numerical integration. The drain current,

$$I_D = \frac{2q k_B T_L}{h} \{F_0(\eta_F) - F_0(\eta_F - U_D)\},$$ (5.57b)

is independent of bandstructure, so it is unchanged from Eqn. (5.6).

Equations (5.57a) and (5.57b) describe the I-V characteristics of ballistic, MOSFET-like CNTFETs. In comparison to a silicon nanowire, we expect better transport (because transport in carbon nanotubes can be near ballistic) and higher on-current through the use of high-k dielectrics, which can be readily deposited, on carbon nanotubes [5.3] but not so readily on silicon. The combination of high-k gate dielectrics with the low, one-dimensional density-of-states, means that it is possible for CNTFETs to operate in the quantum capacitance limit.

5.7 Schottky Barrier Carbon Nanotube FETs

Most present-day CNTFETs operate as Schottky barrier transistors. In modeling such devices, a simple, analytical treatment is not possible, so numerical simulations are essential [5.21]. For a mid-gap Schottky barrier (i.e. one for which the barriers to the conduction and valence bands are equal), the typically observed $\log(I_D)$ vs. V_{GS} characteristic is sketched in Fig. 5.20. This is an n-channel device, but note that a negative gate voltage does not turn the device off. The device displays so-called *ambipolar conduction*. The minimum current occurs at a gate voltage of one-half of the drain voltage. We seek a qualitative understanding of why ambipolar conduction occurs and why the minimum current occurs at $V_G = V_{DD}/2$.

To understand the origin of the ambipolar conduction, consider first the case where $V_{DS} = 0$, so $I_D = 0$. Figure 5.21a shows the band diagram for $V_G = V_D = 0$. For $V_G > 0$, Fig. 5.21b shows that the conduction band is pushed down and electrons tunnel into the conduction band from both contacts. (The net current is zero in this case because there the tunneling is equal and opposite in both directions. For $V_G < 0$, the bands are pulled up, and holes can tunnel in from each contact. The point is that when $V_{GS} > 0$, the conduction band barrier is thinned, and the tunneling current increases. When $V_{GS} < 0$, the valence band barrier is thinned, and the hole tunneling current increases.

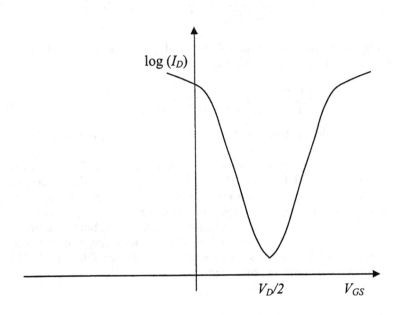

log (I_D)

$V_D/2$ V_{GS}

Figure 5.20   Sketch of the log I_D vs. V_{GS} characteristic of a Schottky barrier CNTFET at $V_D = V_{DD}$.

Consider now the case where $V_G = V_D/2$ as shown in Fig. 5.22. In this case the band diagram is symmetrical from left to right. At the left, $V_G = V_D/2$ and the left contact acts as a source of electrons. At the right contact, $V_G = -V_D/2$ and the right contact acts as a source of holes. Holes and electrons flow in opposite directions, so the two currents add. The minimum current in Fig. 5.20, therefore, consists of equal electron and hole currents. For $V_G > V_D/2$, the device operate as an n-channel FET, and for $V_G < V_D/2$, it operates as a p-channel FET.

For potential applications in CMOS logic, the ambipolar characteristics are undesirable, so the question of how to suppress ambipolar conduction arises. One possibility is to use a metal with $\phi_{Bn} \approx 0$, because the high tunneling barrier for holes ($\phi_{Bp} \approx E_G$) will suppress hole tunneling.

a)

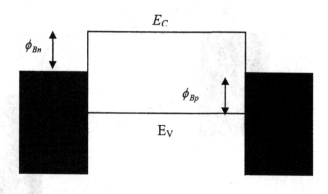

b)

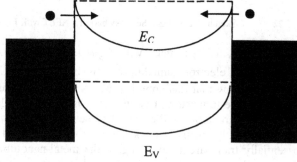

c)

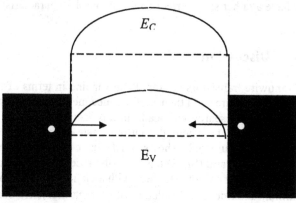

Figure 5.21   Illustration of how both electrons and holes can be injected into an intrinsic nanotube channel depending on whether $V_G > 0$ or $V_G < 0$. V_{DS} = 0 in each case. (a) $V_G = 0$, (b) $V_G > 0$, (c) $V_G < 0$.

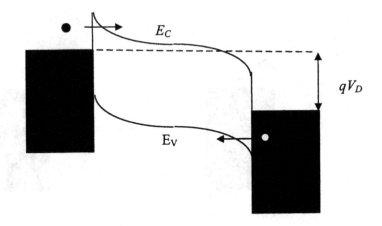

Figure 5.22   Sketch of a mid-gap Schottky barrier CNFET with $V_G = V_{DD}/2$.

Alternatively, using a metal with $\phi_{Bp} \approx 0$ should promote hole current and suppress the electron tunneling current because $\phi_{Bn} \approx E_G$. Numerical studies [5.19] show that this approach works -- when the insulator is thick. High performance transistors, however, will have thin insulators and two-dimensional electrostatics will ensure that the width of the tunneling barrier is approximately the insulator thickness. In that case, the tunneling barriers are essentially transparent, so changing the metal-nanotube barrier height will have a rather small effect on the ambipolar characteristics [5.19].

5.8   Discussion

Nanowire transistors are rapidly advancing in terms of the sophistication of device structures and their performance metrics. At this time, it is hard to say where the research will lead. In this section, we will review one recent result, to illustrate where the current state-of-the-art in carbon nanotube FETs lies. Figure 5.23 shows a schematic cross section and a scanning electron micrograph (SEM) top view of a recently reported CNTFET [5.22]. The device has a carbon nanotube with a diameter of ~1.7 nm ($E_G \sim 0.5$ eV), a hafnium dioxide gate insulator with a high dielectric constant of $\kappa \sim 16$, and a metal gate that is self-aligned to the Pd source and drain contacts. Pd provides a low Schottky barrier to the valence band [5.23].

Figure 5.24 shows the measured I_D vs. V_{DS} characteristics of this device. For a power supply voltage of $V_{DD} = 0.4$V, the on-current (at $V_{DS} = V_{GS} = 0.4$V) is quite high (~12 μA). The obvious questions is: how does this compare to a MOSFET? Comparing the two transistors must be done very carefully, because comparisons are only meaningful if the threshold voltages (or leakage currents) are the same. More difficult is the fact that MOSFET currents are usually quoted per unit micrometer of width, because the current is proportional to W. Currents per unit width are, however, not relevant for nanowire transistors. When careful comparisons are done, the conclusion is that CNTFETs can provide significantly lower device delay at the same I_{ON}/I_{OFF} ratio [5.24].

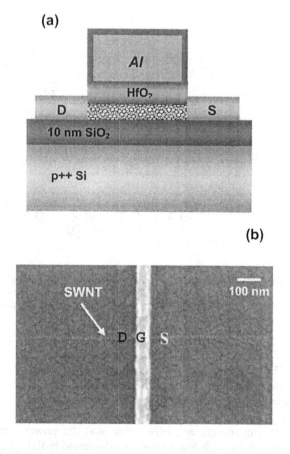

Figure 5.23   Recently reported, high-performance CNTFET (a) Schematic cross-sectional sketch and (b) SEM top view. (Reproduced with permission from [5.22])

Although Pd produces a barrier height of nearly zero to the valence band, the on-current of this device is still limited by quantum mechanical tunneling [5.21]. Figure 5.24 compares the measure I_D vs. V_{DS} characteristic at $V_{GS} = 0.4$V with a simulated characteristic that assumes ballistic transport [5.24]. The comparison suggests that this device operates very near its ballistic limit. Earlier in this chapter, we developed a theory for CNT MOSFETs (CNT SBFETs require a numerical solution.) Also plotted in Fig. 5.24, is the characteristic that a corresponding CNTFET would achieve. The comparison demonstrates that if CNT MOSFETs could be realized, they could deliver about twice as much current as this CNT SBFET.

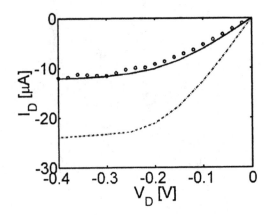

Figure 5.24   Comparison of the measured (circles) and simulated characteristics (solid) at V_G=-0.4V for the CNTFET shown in Fig. 5.23. Also shown is the corresponding result that a CNT MOSFET would achieve (dashed). (Reproduced with permission from [5.24])

The performance of nanowire/nanotube transistors has often been assessed by only quoting the on-current in literature. The assessment, however, is only meaningful when the power supply voltage and the off-current are defined. The performance can be assessed by simultaneously considering on-current, off-current, and the power supply voltage, and a procedure for such comparison was discussed in Ref. [5.24]

5.9   Summary

In this chapter discussed a simple theory of ballistic nanowire MOSFETs. In comparison to traditional MOSFETs with 2D channels, nanowire MOSFETs can display some distinct characteristics, such as a drain conductance that is independent of gate voltage and, in the quantum capacitance limit, a transconductance that is equal to the drain conductance. Because of the low density of states and the ability to make use of high-k gate dielectrics, the quantum capacitance can play an important role in carbon nanotube FETs.

Our scientific understanding and the technological performance of nanowire FETs are progressing rapidly, but many open questions and issues remain. The role of scattering in these devices is one issue. For low biases, symmetry arguments show that elastic backscattering is suppressed, so mean free paths of several hundred nanometers are possible, but under high bias, optical phonon emission occurs and limits the current in a long nanotube to about 25 µA [5.25]. We are just beginning to understand scattering in short nanotubes [5.26, 5.27] and in CNTFETs [5.24]. In addition to potential applications in digital circuits, other possibilities are also of interest. For example, carbon nanotubes have a direct bandgap, so electrically-pumped optical emitters are a possibility [5.28]. Because of their excellent transport properties, CNTFETs are interesting as high-frequency transistors [5.29]. The possibility of using nanowires to achieve room temperature single electron transistors is also interesting [5.30]. Finally, nanowires are interesting as high-sensitivity sensors for bio-medical applications [5.31, 5.13]. Where this will all lead is hard to predict, but we are certain to learn more about the physics of small electronic devices as we explore these possibilities.

Chapter 5 References

[5.1]   S.J. Tans, R.M. Verschueren, and C. Dekker, "Room Temperature Transistor Based on a Single Carbon Nanotube," *Nature,* **393**, pp. 49-52, 1998.

[5.2]   R. Martel, T. Schmidt, H.R. Shea, T. Hertel, and Ph. Avouris, "Single and Multi-Wall Carbon Nanotube Field-Effect Transistors, *Appl. Phys. Lett.,* **73**, pp. 2447-2449, 1998.

[5.3]   A. Javey, H. Kim' M. Brink, Q. Wang, A. Ural, J. Guo, P. McIntyre, P. McEuen, M. Lundstrom, and H. Dai, "High   Dielectrics For Advanced Carbon Nanotube Transistors and Logic," **1**, pp. 241-246 *Nature Materials,* 2002.

[5.4]   C. M. Lieber, "The Incredible shrinking Circuit" *Scientific American,* **285**, pp. 58-64, 2001

[5.5]   Y. Cui, Z. Zhong, D. Wang, W. U. Wang, and C.M. Lieber, "High Performance Silicon Nanowire Field effect Transistors," *Nano Letters,* **3**, pp. 149-153, 2003.

[5.6]   S. Iijima and T. Ichihashi, "Single-Shell Carbon Nanotubes of 1-nm Diameter," *Nature,* **363**, pp. 603-605, 1993.

[5.7]   D.S. Bethune, C.H. Kiang, M.S. Devries, G. Gorman, R. Savoy, J. Vazquez, and R. Beyers, "Cobalt-Catalyzed Growth of Carbon Nanotubes with Single-Atomic-Layerwalls," *Nature,* **363**, pp. 605-607, 1993

[5.8]   R. Saito, G. Dresselhaus, and M.S. Dresselhaus, *Physical Properties of Carbon Nanotubes,* Imperical College Press, London, 1998.

[5.9]   P. McEuen, M. Fuhrer, and H. Park, "Single-Walled Nanotube Electronics," *IEEE Trans. on Nanotechnology,* 1, pp. 78-85, 2002.

[5.10]   A. Bachtold, P. Hadley, T. Nakanishi, and C. Dekker, "Logic Circuits with Carbon Nanotube Transistors," *Science,* **294**, pp. 1317-1320, 2001.

[5.11]   V. Derycke, R. Martel, J. Appenzeller, and P. Avouris, "Carbon Nanotube Inter- and Intramolecular Logic Gates," *Nano Letters,* **1**, pp. 453-456, 2001.

[5.12]   J. Kong, J. Cao, and H. Dai, "Chemical Profiling of Single Nanotubes: Intramolecular p-n-p junctions and on-tube single-electron transistors," *Appl. Phys. Lett.,* **80**, p. 73, 2002.

[5.13]   R.J. Chen, S. Bangsaruntip, K.A. Drouvalakis, N. W.S. Kam, M/ Shim, Y. Li, W. Kim, P.J. Utz, and H. Dai, "Noncovalent functionalization of Carbon Nanotubes for Highly Specific Electronics Biosensors," *PNAS,* **100**, pp. 4984-4989, 2003.

[5.14]   J.A. Misewich, R. Martel, Ph. Avouris, J.C. Tsang, S. Heinze, and J. Tersoff, "Electrically Induced Optical Emission from a Carbon Nanotube FET," *Science,* 2003.

[5.15]   L. Yang, M. P. Anantram, and J. P. Lu, "Band-gap change of carbon nanotubes: Effect of small uniaxial and torsional strain," Physical Review B., vol. 60, no. 29, pp. 13874-13878, 1999.

[5.16]   Supriyo Datta, *Quantum Transport: from Atom to Transistor,* in press Cambridge University Press, Cambridge, UK, 2004.

[5.17]   J. W. Mintmire and C. T. White, *Phys. Rev. Lett.*, "Universal Density of States for Carbon Nanotubes," **81**, p. 2506, 1998.

[5.18]   M.P. Lepselter and S.M. Sze, "SB-IGFET: An insulated gate field-effect transistor using schottky barrier contacts as source and drain," *Proc. IEEE*, **56**, p. 1088, 1968.

[5.19]   J. Guo, S. Datta, and M.S. Lundstrom, "A numerical study of scaling issues for Schottky barrier carbon nanotube transistors," *IEEE Trans. on Electron Devices*, **51**, pp. 172-177, 2004.

[5.20]   J. Appenzeller, et al., "Field modulated carrier transport in carbon-nanotube transistors," *Phys. Rev. Lett.*, **89**, p. 126801, 2002.

[5.21]   A. Javey, R. Tu, D. B. Farmer, J. Guo, R. Gordon, H. Dai, "High Performance n-Type Carbon Nanotube Field-Effect Transistors with Chemically Doped Contacts" *Nano Lett.*, **5**, p. 345, 2005.

[5.22]   A. Javey, J. Guo, D. B. Farmer, Q. Wang, E. Yenilmez, R. G. Gordon, M. Lundstrom, and H. Dai, "Self-aligned ballistic molecular transistors and parallel nanotube arrays," *Nano Lett.*, **4**, pp. 1319-1322, 2004.

[5.23]   A. Javey, J. Guo, Q. Wang, M. Lundstrom, and H. Dai, ""Ballistic Carbon Nanotube Field-Effect Transistors," *Nature*, **424**, pp. 654-657, 2003. (See also, News and Views by J. Tersoff, "Nanotechnology: A Barrier Falls," p. 622).

[5.24]   J. Guo, A. Javey, H. Dai, and M. Lundstrom "Performance Analysis and Design Optimization of Near Ballistic Carbon Nanotube Field-Effect Transistors," *International Electron Devices Meeting Tech. Digest (IEDM)*, pp. 703-706 San Francisco, CA, Dec. 13-15, 2004

[5.25]   Z. Yao, C. L. Kane, and C. Dekker, "High-field electrical transport in single-wall carbon nanotubes," *Phys. Rev. Lett.*, **84**, pp. 2941-2944, 2000.

[5.26]   A. Javey, J. Guo, M. Paulsson, Q. Wang, D. Mann, M. Lundstrom, and H. Dai, "High-Field Quasi-Ballistic Transport in Short Carbon Nanotubes," *Phys. Rev. Lett.* **92**, p. 106804, 2004.

[5.27]   J. Y. Park, S. Rosenblatt, Y. Yaish, V. Sazonova, H. Ustunel, S. Braig, T. A. Arias, P. W. Brouwer, and P. L. McEuen, "Electron-phonon scattering in metallic single-walled carbon nanotubes," *Nano Letters*, **4**, pp. 517-520, 2004.

[5.28]   J. A. Misewich, R. Martel, Ph. Avouris, J. C. Tsang, S. Heinze, and J. Tersoff, "Electrically induced optical emission from a carbon nanotube FET," *Science*, **300**, pp. 783-786, 2003.

[5.29]   J. Appenzeller and D. J. Frank, "Frequency dependent characterization of transport properties in carbon nanotube transistors," *Appl. Phys. Lett.*, **84**, pp. 1771-1773, 2004.

[5.30]   H.W. Ch. Postma, T. Teepen, Z. Yao, M. Grifoni, and C. Dekker, "Carbon nanotube single-electron transistors at room temperature," *Science*, **293**, 76-79, 2001.

[5.31]   Y. Cui, Q. Q. Wei, H. K. Park, and C. M. Lieber, "Nanowire nanosensors for highly sensitive and selective detection of biological and chemical species," *Science*, vol. 293, pp. 1289-1292, 2001

Chapter 6:   Transistors at the Molecular Scale

6.1 Introduction

In this monograph, we began by considering the traditional MOSFET, which has a two-dimensional channel.   We then discussed nanowire transistors, which have one-dimensional channels. To conclude, we should consider the possibility of a zero-dimensional channel, which would occur, for example, if we could make a transistor out of a single molecule. Before considering molecular transistors, however, we need to understand how molecules conduct electricity. Conduction in molecules is still an active topic of research, but an understanding is beginning to emerge, and a simple conceptual picture explains a lot [6.1].   This conceptual picture of conduction in molecules can be generalized into a theory for ballistic nanotransistors that applies to transistors with channels of any dimensionality [6.1]. The simple model demonstrates that single molecule transistors are possible, and numerical simulations can be used to identify the challenges quantitatively.   Finally, we must address single electron charging.   As transistor channels shrink, the number of electrons in the channel decreases. It is possible to have even less than one electron in the channel, because it is spatially extended over the source, channel, and drain. A special kind of transistor known as a single electron transistor can also be realized. In a single electron transistor, only a discrete number of electrons can exist in the channel, and the number can be controlled by the gate voltage. They are not a replacement for the MOSFET, but their unique characteristics are potentially useful in conjunction with CMOS. The "general theory" that we have discussed in this monograph does not apply to single electron transistors. We will explain why a special treatment is necessary.

6.2 Electronic Conduction in Molecules

In recent years, experimentalists have devised several ways to measure the *I-V* characteristics of an individual molecule (or a small number of molecules in parallel) [6.2-6.7]. One possible experimental situation is illustrated in Fig. 6.1, which shows a small organic molecule, with a sulfur atom attached to one end. Sulfur forms a strong bond with gold, which is used to attach the molecule to a gold substrate. The second contact in this illustration is the tip of a scanning tunneling microscope. Several different types of *I-V* characteristics are observed, including switching [6.4] and negative differential resistance [6.5]. These complex *I-V* characteristics are not completely understood, and some of them may involve oxidation-reduction processes and a change in the conformation (shape) of the molecule. In the simplest case, however, the *I-V* characteristics resemble the sketch in Fig. 6.2a, which suggests two back-to-back diodes. As shown in Fig. 6.2b, there is a conductance gap, where $G \approx 0$, and for voltages above the conductance gap, the magnitude of the current increases with bias. Characteristics like these are thought to involve molecules that are more or less rigid. For these kinds of experiments, an ability to simulate the *I-V* characteristic and to understand it in term of a simple, conceptual model has been developed [6.1, 6.8, 6.9]. We will use this simple conceptual viewpoint as the basis for a general model for ballistic nanotransistors.

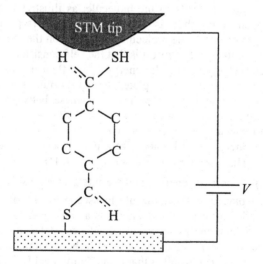

Figure 6.1   Schematic illustration of how the *I-V* characteristics of a single molecule can be measured. (After [6.2])

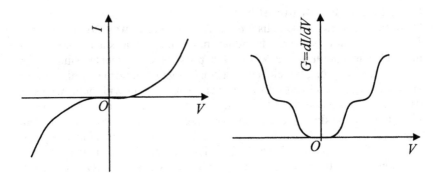

Figure 6.2   A commonly observed current (a) and conductance (b) vs. voltage
              characteristic for molecular conductors. (After [6.8])

As sketched in Fig. 6.3a, an isolated molecule has a set of discrete energy
levels. A discrete number of electrons fill up these energy levels to some
point. The set of highest occupied molecular orbitals (known as HOMO)
are analogous to the valence band of a semiconductor. Above them is a set
of lowest unoccupied molecular orbitals (LUMO). The LUMO are
analogous to the conduction band of a semiconductor.

When we attach contacts to the molecule, as illustrated in Fig. 6.3b, two
things happen. First, there is some charge transfer (typically less than one
electron) into or out of the molecule depending on the location of the Fermi
level in the contacts. The result is that the self-consistent potential, U_{SCF}, in
the molecule changes, and the energy levels float up or down. After the
charge transfer process is complete, and equilibrium is achieved, the Fermi
level is constant and typically lies somewhere between the HOMO and
LUMO, (we have shown it nearer to the HOMO in this sketch). When we
connect a molecule to the outside world, there is a second effect that also
occurs; the energy level broaden from their discrete values of the isolated
molecule. The broadening is given by $\gamma = \hbar \left(1/\tau_1 + 1/\tau_2 \right)$, where γ is the
broadening in units of energy, and we interpret $\tau_{1,2}$ as the time it takes for
an electron placed in the molecule to escape to one of the two contacts.
(Level broadening can be understood as a consequence of the Uncertainty
Principle.) When energy levels are broadened, states are conserved, so one
(times two for spin) electrons can still be accommodated in a single
broadened level. The result is that if the Fermi level lies inside a broadened
level, then the molecule can hold a fractional number of electrons.

Now consider what happens under bias. Assume that the left contact is grounded ($V = 0$) and that the bias is applied to the right contact as shown in Fig. 6.3c. How the molecular levels move will be determined by how well the molecule is connected to the two contacts. If the molecule is well connected to the left contact but poorly connected to the right contact, then the levels will move little with respect to the potential on the left contact, and most of the voltage drop will occur at the right contact. The opposite will happen if the molecule is well connected to the left contact and poorly connected to the right contact. If the molecule is connected equally well to both contacts, then the applied bias should drop equally at the two contacts.

Assume a positive bias is applied to the right contact and that the molecule is symmetrical with respect to the two contacts. No current flows until the Fermi level of the right contact moves below the HOMO level, which requires a voltage of $V = 2\left(E_F - \varepsilon_{HOMO}\right)/q$ at the right contact. (The factor of two occurs because half of the voltage drop occurs at the each contact.) Electrons flow from the left to right (which we define as a positive current). The Fermi level in the left contact lies above the filled levels in the molecule, so the left contact wants to put electrons into the molecule. At the same time, the Fermi level in the right contact lies below the filled levels, so electrons in the molecule want to flow out into the right contact. Note that conduction is through a filled level in this case, which is analogous to the conduction through the valence band in a semiconductor.

Consider now the application of a negative bias to the right contact as in Fig 6.3d. The negative bias raises the Fermi level, but no current flows until $V < -2\left(E_F - \varepsilon_{HOMO}\right)/q$ so that the HOMO lies between the two Fermi levels. In this case, the current flows in the opposite direction because the right contact wants to put electrons into the molecule, and the left contact wants to pull them out. Conduction, however, is still through the filled levels, the HOMO.

This simple conceptual picture of conduction in a molecule explains the commonly observed *I-V* characteristics of molecules as sketched in Fig. 6.1 and is confirmed by more detailed numerical simulations [6.10, 6.11, 6.12]. Note that the magnitude of the conductance gap is related to the separation of the Fermi level and the nearest level, whether it is filled or empty. The conductance gap is not the HOMO to LUMO gap, as might have been expected. It we were to repeat the exercise assuming that the equilibrium Fermi level lies near the LUMO, we would find a similar *I-V* characteristic, although conduction would occur through an empty level in that case. It is not possible to tell from the measured *I-V* characteristic whether conduction is through the HOMO or LUMO. This simple conceptual view seems to

explain much of the available data, and a prediction based on it, that if one
of the two contacts is replaced by a semiconductor, the negative differential
resistance will occur for one bias polarity but not the other [6.1] has been
confirmed experimentally [6.7]. We will use this conceptual picture to
develop a general model for the kinds of nanotransistors that we have been
discussing in this monograph.

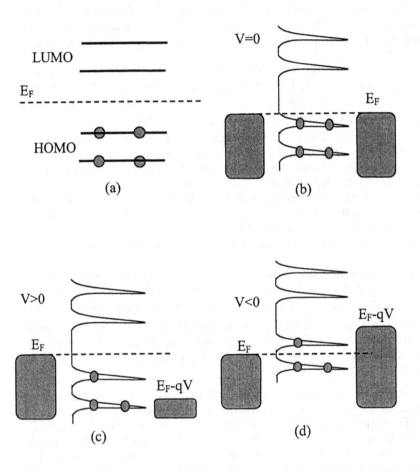

Figure 6.3   Simple conceptual view of conduction in a molecule (a) The energy levels of
an isolate molecule. (b) The broadened and shifted energy levels of a
molecule connected to two contacts. (c) The molecule with a positive voltage
applied to the right contact. (d) The molecule with a negative voltage applied
to the right contact. (After [6.11])

6.3 General Model for Ballistic Nanotransistors

Figure 6.4 illustrates the general model that we will use for ballistic nanotransistors. The "device" may be a single molecule, a nanowire, or the channel of a conventional MOSFET. It is connected to two contacts, and the strength of the connection is describe by the parameters,

$$\gamma_{1,2} = \frac{\hbar}{\tau_{1,2}}. \qquad (6.1)$$

The parameter, γ, in units of energy, represents the broadening of the energy levels in the device due to their interaction with the outside world, and τ represents the time it takes for an electron placed in the device to escape to one of the two contacts. The device is connected to two reservoirs that are assumed to be in thermal equilibrium and represented by their respective Fermi levels, E_{F1} and E_{F2}. The device itself is described by a density-of-states, $D(E - U_{SCF})$ and a self-consistent potential, U_{SCF}. In general, we should spatially resolve the density-of-states and self-consistent potential within the device, but for the simple (almost analytical) model that we seek, we will assume a spatially constant density-of-states and self-consistent potential within the device.

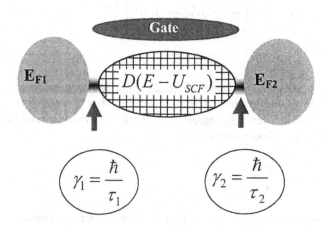

Figure 6.4   A simple conceptual representation of a nanotransistor.

Our model for the nanotransistor is similar to our model for a molecular conductor, except for the presence of a third terminal, the gate electrode. The two contacts still seek to fill the states in the device according to their Fermi levels, but the self-consistent potential, which moves the energy levels up and down, is now controlled by the potential on all three electrodes. A good transistor, however, is designed so that the self-consistent potential within the device is mostly controlled by the gate voltage and is affected very little by the source and drain voltages. To formulate a model for nanotransistors, we need to consider two things – electrostatics and transport. Electrostatics determines how the self-consistent potential is related to the applied voltages, and transport determines how the levels are filled from the two contacts. The two problems are coupled, so a self-consistent solution will be necessary.

Electrostatics of nanotransistors

Consider electrostatics first. We can represent the electrostatic problem by the capacitor circuit shown in Fig. 6.5a (which is the same one we used in Chapter 4). Note that the self-consistent potential in the device is determined by the voltages applied to the terminals <u>and</u> by the charge in the device. We can solve this problem by superposition. First, assume that the charge is zero, which allows us to determine the potential in the device by simple voltage division among the capacitors as sketched in Fig. 6.5b. The result is the Laplace (zero charge) solution

$$U_L = -qV_G\left(\frac{C_G}{C_\Sigma}\right) - qV_D\left(\frac{C_D}{C_\Sigma}\right) - qV_S\left(\frac{C_S}{C_\Sigma}\right), \tag{6.2}$$

where

$$C_\Sigma = C_G + C_D + C_S. \tag{6.3}$$

Next, we assume that the terminals are grounded and place a charge, $Q = -qN$ C, in the device. (Here, N is the number of electrons in the device, which is not necessarily discrete.) From the equivalent circuit of Fig. 6.5c, we find the Poisson solution as

$$U_P = \frac{q^2 N}{C_\Sigma}. \tag{6.4}$$

Finally, the complete solution for the self-consistent potential is

$$U_{SCF} = -qV_G\left(\frac{C_G}{C_\Sigma}\right) - qV_D\left(\frac{C_D}{C_\Sigma}\right) - qV_S\left(\frac{C_S}{C_\Sigma}\right) + \frac{q^2 N(U_{SCF})}{C_\Sigma}. \tag{6.5}$$

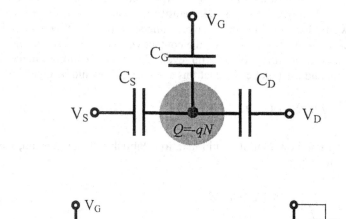

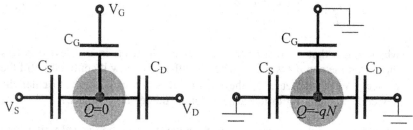

Figure 6.5   A capacitor model to represent the electrostatics of the nanotransistor. (a) The general case with charge in the device and voltages applied to the terminals. (b) The case with no charge in the device. (c). The case with charge in the device, but no voltages applied.

Equation (6.5) shows that a positive gate voltage tries to push the energy levels down, but the presence of charge in the device causes them to float up. For an electrostatically well-designed MOSFET, $C_G \approx C_\Sigma$ and $C_D, C_S \ll C_\Sigma$. As Eqn. (6.5) indicates, the number of electrons in the device itself depends on the self-consistent potential, because it moves the energy levels up and down with respect to the two Fermi levels, so we must now specify how the states in the device are filled from the two contacts.

Transport in ballistic nanotransistors

From the discussion in Sec. 6.2, we know that the number of electrons in the device is determined by a competition between what the two contacts want it to be. We now need to quantify this observation. Consider first what would happen if the device were connected to the left contact only, as shown in Fig. 6.6a. In this case, the contact would want to fill up the states in the device according to its Fermi level. Eventually, equilibrium would be achieved, and the number of electrons at energy, E, would be given by

$$N_1^o(E) = D(E - U_{SCF})f_1(E).$$ (6.6)

To determine how long it would take to establish this equilibrium, we can write a rate equation,

$$\frac{dN(E)}{dt} = \frac{N_1^o(E) - N(E)}{\tau_1},$$ (6.7)

which states that as long as the number of electrons in the device is less than what the contact wants it to be, then the contact will continue to fill the device. We interpret τ_1 as the time it takes for an electron to enter the device from contact 1 or to escape from the device into contact 1.

Consider next what would happen if the device were connected to the right contact only, as shown in Fig. 6.6b. The right contact would want to fill the device up according to its Fermi level. When equilibrium is achieved, the number of electrons at energy, E, would be given by

$$N_2^o(E) = D(E - U_{SCF})f_2(E).$$ (6.8)

The rate at which contact 2 would fill the device is given by

$$\frac{dN(E)}{dt} = \frac{N_2^o(E) - N(E)}{\tau_2}.$$ (6.9)

In general, the device is connected to both contacts, and each one wants to fill it up according to its own Fermi level. We have, therefore,

$$\frac{dN(E)}{dt} = \frac{N_1^o(E) - N(E)}{\tau_1} + \frac{N_2^o(E) - N(E)}{\tau_2}. \tag{6.10}$$

In steady-state, $dN/dt = 0$, so we can solve Eqn. (6.10) for the number of electrons in the device under steady-state conditions to find

$$N(E - U_{SCF}) = D_1(E - U_{SCF})f_1(E) + D_2(E - U_{SCF})f_2(E), \tag{6.11}$$

where

$$D_{1,2}(E - U_{SCF}) = \frac{\gamma_{1,2}}{\gamma_1 + \gamma_2} D(E - U_{SCF}) \tag{6.12}$$

is the density-of-states filled from each of the two contacts. If the device is equally well connected to both contacts, then half of the states are filled by each contact. In our ballistic model, the energy channels are independent, so the total number of electrons in the device is given by

$$N(U_{SCF}) = \int dE[D_1(E - U_{SCF})f_1(E) + D_2(E - U_{SCF})f_2(E)]. \tag{6.13}$$

Equation (6.13) shows that N is a function of the self-consistent potential, and Eqn. (6.5) shows that U_{SCF} is a function of N, so these two equations need to be solved iteratively to find a self-consistent solution.

After evaluating U_{SCF} and N, we can find the steady-state current. Under steady-state conditions, one contact puts electrons into the device at the same rate that the other contact removes them. From Eqn. (6.7) and (6.9), we find the current in an energy channel as

$$I(E) = \frac{q(N_1^o - N)}{\tau_1} = -\frac{q(N_2^o - N)}{\tau_2}. \tag{6.14}$$

Using Eqn, (6.6) and (6.11), we find

$$I(E) = \frac{2q}{h} T(E - U_{SCF})[f_1(E) - f_2(E)], \tag{6.15}$$

where

$$T(E - U_{SCF}) = 2\pi \frac{\gamma_1 \gamma_2}{\gamma_1 + \gamma_2} D(E - U_{SCF}) \qquad (6.16)$$

is the current transmission at energy, E. The total current is found by summing over all the energy channels,

$$I_D = \int dE\, I(E) = \frac{2q}{h} \int dE\, T(E)\left[f_1(E) - f_2(E)\right]. \qquad (6.17)$$

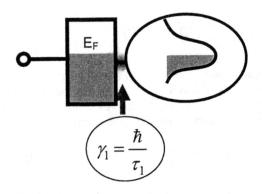

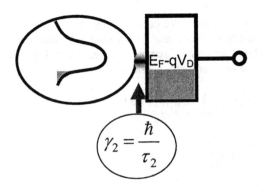

Figure 6.6   Illustration of how the two contacts try to fill the device according to their own Fermi levels. (a) The device connected to the left contact only and (b) the device connected to the right contact only.

Summary
The general model for ballistic nanotransistors can be summarized as follows. Given a device structure, we first estimate the three capacitors, C_G, C_D, and C_S. Second, we specify the connection to the reservoir, so that the two escape times, τ_1, and τ_2, (or the broadenings, γ_1, and γ_2) can be evaluated. Then we assume a density-of-states for the device (a more general model would compute the locate density-of-states self-consistently). Finally, we specify the value of the two Fermi levels.

Having specified the device, we compute its *I-V* characteristics in two steps. First, we solve the electrostatic problem,

$$N = f_N\left(U_{SCF}, V_D, V_S\right) \tag{6.18a}$$

$$U_{SCF} = f_U\left(V_G, V_D, V_S, N\right), \tag{6.18b}$$

where the functions, f_N and f_U are specified by Eqns. (6.13) and (6.5). Because they are coupled, we solve them iteratively to obtain a self-consistent solution. After obtaining U_{SCF}, we compute the current

$$I_D = f_I\left(U_{SCF}, V_D, V_S\right), \tag{6.18c}$$

from Eqn. (6.17). In the next section, we apply this procedure to three different kinds of transistors.

6.4 MOSFETs with 0D, 1D, and 2D Channels

To model a ballistic MOSFET, we should first relate the electron number in the device to the self-consistent potential within the device, U_{SCF}, by specifying the relationship of Eqn. (6.18a). The equation for the self-consistent potential, Eqn. (6.18b), is then specified. Finally we specify the functional relationship, Eqn. (6.18c), which defines the current. In this section, we discuss each of these three steps.

The electron density
To keep the discussion simple, we assume that $\gamma_1 = \gamma_2 = \gamma$, so that Eqn. (6.13) for the electron number becomes

$$N\left(U_{SCF}, E_{F1}, E_{F2}\right) = \int \frac{D\left(E - U_{SCF}\right)}{2}\left[f_1(E) + f_2(E)\right]dE. \tag{6.19}$$

We consider a molecular transistor to be a quasi-0D device with a set of broadened molecular levels. A nanowire transistor has a quasi-1D electronic structure, and a traditional MOSFET has a quasi-2D electronic structure. To evaluate Eqn. (6.19), we simply need to specify the appropriate density of states.

To keep the mathematics simple, we will assume conduction through a single level or subband (the generalization to multiple levels or subbands is straightforward). Figure 6.7 sketches the three density-of-states that will be assumed. For the molecular transistor, we assume that the broadened level is described by a Lorentzian,

$$D_{0D}(E - U_{SCF}) = 2 \times \frac{\gamma/2\pi}{(E - U_{SCF})^2 + (\gamma/2)^2}, \tag{6.20a}$$

where the factor of 2 in front accounts for spin degeneracy. When the Lorentzian density of states is integrated over all energies, the result is 2, because the broadened level can hold the same number of electrons as the initial discrete level.

For a nanowire transistor with a simple density of states from a parabolic bandstructure, the density of states is given by Eqn. (1.16) as

$$D_{1D}(E - U_{SCF}) = L\left(\frac{\sqrt{2m^*}}{\pi\hbar}\right)\frac{1}{\sqrt{E - U_{SCF}}}\Theta(E - U_{SCF}), \tag{6.20b}$$

where Θ is the unit step function and L is the "length" of the active part of the device. (Recall that the density of states is expressed in units of number of states per energy, not number per unity length per energy as is common.) For carbon nanotube MOSFETs, we use a slight different density of states as given by Eqn. (5.52),

$$D_{CNT}(E - U_{SCF}) = LD_0 \frac{E}{\sqrt{E^2 - U_{SCF}^2}}\Theta(E - U_{SCF}), \tag{6.20c}$$

where D_0 is the metallic density of states as given by Eqn. (5.48).
Finally, for a traditional MOSFET with simple, parabolic energy bands, we have a constant density of states,

$$D_{2D}(E - U_{SCF}) = WL\frac{m^*}{\pi\hbar^2}\Theta(E - U_{SCF}). \tag{6.20d}$$

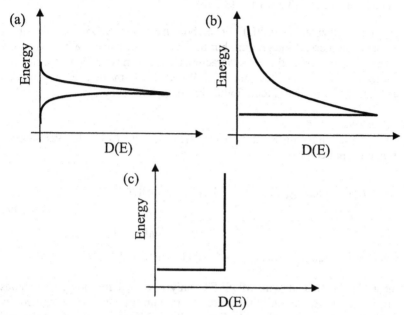

Figure 6.7   Illustration of the assumed density of states for (a) a single level molecular
              MOSFET (0D), (b) a nanowire or nanotube MOSFET (1D), and (c) a
              traditional MOSFET with a 2D channel.

Having specified the density of states, we can integrate Eqn. (6.19). For the
semiconductor nanowire or traditional MOSFET with simple, parabolic
energy bands, we can express the result in terms of tabulated integrals. For
a semiconductor nanowire, MOSFET, we find

$$N(U_{SCF}, E_{F1}, E_{F2}) = \frac{L}{2} \sqrt{\frac{2m^{*}k_{B}T_{L}}{\pi\hbar^{2}}} [\Im_{-1/2}(\eta_{F1}) + \Im_{-1/2}(\eta_{F2})], \quad (6.21a)$$

where $\eta_{F1,F2} = (E_{F1,F2} - U_{SCF})/k_{B}T_{L}$. Equation (6.21a) is simply Eqns.
(5.4) that we saw in Chapter 5 but expressed in electron number rather than
in number density.

Similarly, for the traditional MOSFET, we find

$$N(U_{SCF}, E_{F1}, E_{F2}) = \frac{WL}{2} \frac{m^{*}k_{B}T_{L}}{\pi\hbar^{2}} [\Im_{0}(\eta_{F1}) + \Im_{0}(\eta_{F2})], \quad (6.21b)$$

which is just Eqns. (3.33a) and (3.33b).

For the molecular MOSFET or carbon nanotube MOSFET, we must perform a numerical integration, but the result is the same. When writing a program to compute the *I-V* characteristics, we must call a routine that returns N given U_{SCF}, E_{F1}, and E_{F2}. Whether we make use of tabulated functions or numerical integration makes little difference.

The current
For one subband with ideal contacts, $T(E)=1$, Eqn. (6.17) for the source-drain current becomes

$$I(U_{SCF}, E_{F1}, E_{F2}) = \frac{2q}{h} \int_{U_{SCF}}^{+\infty} [f_1(E) - f_2(E)]dE$$
$$= \frac{2qk_BT}{h}[\Im_0(\eta_{F1}) - \Im_0(\eta_{F2})]$$

(6.22)

which is just Eqns. (5.4c) and (5,44d) that we saw in Chapter 5.

For a CNT or a nanowire MOSFET, only one or a few lowest subbands carries the current, and the source-drain current can be computed by summation over these subbands. For a ballistic MOSFET with a 2D channel, the source-drain current can be computed by summation over all transverse modes, similar to what we did in section 3.4.2 for the channel conductance.

6.5 Molecular Transistors?

The simple model as described above provides many insights into how molecular transistors operate. In this section, we discuss the possibility and scaling issues of molecular transistors using detailed numerical simulations from Ref. 6.11. The quantum effects and transistor electrostatics are rigorously treated by numerically solving the quantum transport equation using the NEGF formalism self-consistently with the Poisson equation.

Simulation results indicate that molecular transistors will have the same scaling issues as conventional transistors [6.11, 6.13]. Electrostatic short channel effects, quantum mechanical tunneling, and variations between transistors still set the scaling limit of this hypothetical molecular transistor, just as they do for a silicon MOSFET. Molecular transistors, therefore, are not fundamentally more scalable than silicon MOSFETs, though small difference of the scaling limit may exist due to different device geometry and material properties.

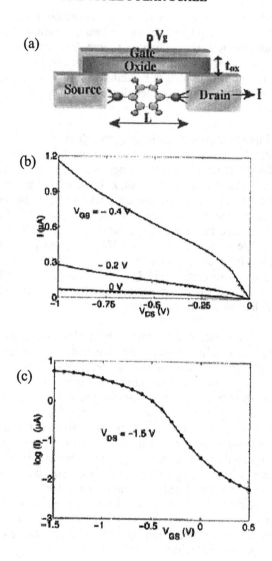

Figure 6.8   Simulated I-V characteristics of a molecular transistor. (a) Device structure
with a channel length, L~1.0nm. (b) I_D-V_D characteristics and (c) $log10(I_D)$-V_G
characteristics. The gate oxide thickness is t_{ox}=0.1nm, and the gate leakage
current is omitted in the simulation. (Reproduced with permission from [6.11])

Figure 6.8a shows the schematic sketch of a molecular transistor. The transistor channel is a single phenyl dithiol molecule, which is attached to the source and drain contacts. By applying a source-drain voltage, a current flows through the molecular channel. The source-drain current is modulated by a third electrode, the gate, which is insulated from the molecule by the gate oxide. The small size of the molecule results in a short transistor channel of $L_{ch} \sim 1nm$.

The short channel length makes it challenging to achieve good electrostatic control by the gate. The source-drain current of a well-designed transistor should be controlled by the applied gate voltage, instead of the source-drain voltage. This requires that the channel length, L, be several times larger than the gate oxide thickness, t_{ox}. Because the channel length of the molecular transistor is so short, the short channel effects are severe. Numerical simulation results indicate that the I_D-V_D curves do not saturate at all if a 1.5nm-thick gate oxide is used [6.11]. Even if the gate oxide thickness is only 0.1nm (which is impractically thin because the oxide needs to be thick enough to behave as an insulator), the I_D-V_D still has a large differential output conductance in the saturation region, as shown in Fig. 6.8b. The electrostatic short channel effects are severe.

In addition to poor electrostatics, quantum mechanical tunneling through the short channel also severely degrades the transistor performance. An important figure of merit for a transistor is the subthreshold swing. As discussed in Chapter 2, for a well-designed MOSFET, the subthreshold swing at room temperature approaches ~60mV/dec, and decreases linearly with $k_B T$ as temperature decreases. As shown in Fig. 6.8c, the subthreshold swing of the simulated molecular transistor is about 300mV/dec and nearly temperature-independent, even if an ideal gate control is assumed and the leakage current through the gate oxide is neglected. The subthreshold swing at room temperature is 5 times larger than a well designed silicon MOSFET, which indicates that the simulated molecular transistor is a poor gate controlled switch. The reason is that the channel length is so short that electrons can directly tunnel from the source to the drain contact. The source-drain tunneling prevents the transistor from quickly turning off and results in a large subthreshold swing.

This molecular transistor operates by raising and lowering energy states with a gate voltage. Fundamentally, it operates just like a silicon MOSFET. The lesson is that molecular transistors of this type do not address the key challenges of ultimate CMOS.

6.6 Single Electron Charging

The coupling strength between the channel and the source (drain) contact plays an important role on the *I-V* characteristics of a generic transistor as shown in Fig. 6.8. When the coupling between the channel and the source (drain) contact is strong, the electron wave function can spatially extend over the source, channel, and drain. The total charge on the island can vary continuously. The general theory for a MOSFET as described in the previous sections applies, and I-V characteristics as shown in Figs. 6.9a and 6.9b are expected. In contrast, if the coupling between the channel and the source and drain contacts is weak, the electron wave function is confined in the channel. The total charge of the channel (island) then varies discretely in steps of a single electron charge, $q = 1.6 \times 10^{-19} C$, and the device becomes a different kind of transistor – a single electron transistor (SET) [6.14-6.17]. The *I-V* characteristics of a SET are qualitatively different, as shown Figs. 6.9c and 6.9d. The I_D-V_G curve shows periodic oscillations called Coulomb oscillations. The I_D-V_D curve shows a blockage region near V_D=0 called Coulomb blockade. Both Coulomb blockade and Coulomb oscillations are due to the discreteness of charge on the island, which is not captured in the general model discussed in 6.3. The model breaks down for a single electron transistor.

Before discussing how to understand the characteristics of a single electron transistor, we first define the conditions for observing the single electron charging effect. The following two conditions must be satisfied [6.15]:

1) The thermal energy, $k_B T$, is much smaller than the single electron charging energy, $U_0 = q^2 / C_\Sigma$, where q is a single electron charge and C_Σ is the total capacitance of the island. To obtain a large enough single electron charging energy, the size of the island needs to be small. For example, to observe single electron charging at room temperature, $k_B T \approx 26 meV$, the radius of a spherical island (with $C_\Sigma = 4\pi\varepsilon R$) must be much smaller than 10nm, $R << q^2 /(4\pi\varepsilon k_B T) \sim 10nm$.

2) The tunneling resistance of the source and drain junctions is much larger than the quantum resistance, $R_{S,D} >> R_Q$, where the quantum resistance $R_Q = h/(2q^2) \approx 13 K\Omega$. In order to observe the single electron charging effect, the electron wave function needs to be confined in the island. This requires the transmission through the junctions T<<1 and the junction resistance $R_{S,D} \sim (1/T)(h/2q^2) >> R_Q$.

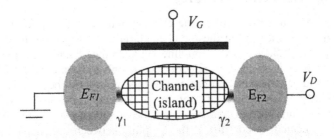

Figure 6.8   A generic transistor. If the coupling between the channel and the source (drain) contact is strong, the transistor is a MOSFET, as described by the general model in the section 6.3. If the coupling is weak, the transistor operates as a single electron transistor, for which the general theory breaks down.

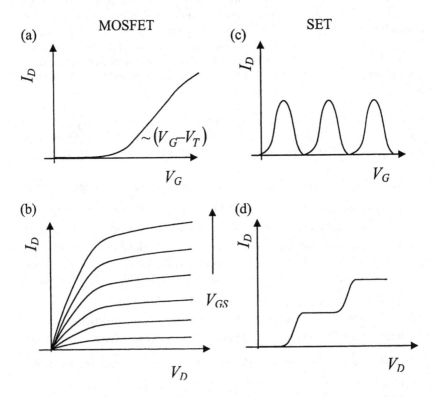

Figure 6.9   The *I-V* characteristics of a MOSFET and a SET are qualitatively different. (a) I_D-V_G and (b) I_D-V_D characteristics for a MOSFET. (c) I_D-V_G and (d) I_D-V_D characteristics for a SET.

Coulomb blockade

Figure 6.10a shows a two terminal single electron device without the gate electrode. A small metal island in this case is weakly coupled to the source and drain contacts. When a drain voltage is applied, electrons can tunnel from and to the island through two tunneling junctions. Identical source and drain tunneling junctions and zero temperature, $T=0$, are assumed for simplicity.

We first plot the equilibrium band profile in Fig. 6.10b. At $T=0$, all states below the Fermi level are filled and those above the Fermi level are empty. Notice that even for a metal island, an energy gap exists between the highest filled state and the lowest empty state due to the single electron charging effect. The single electron charging energy gap is

$$U_0 = q^2 / C_\Sigma,$$ (6.23)

where C_Σ is the total capacitance of the island.

The single electron charging energy gap can be derived as follows. If there are N electrons on the island, the Coulomb interaction energy of the N electrons is

$$E(N) = \frac{q^2 N(N-1)}{2C_\Sigma}.$$ (6.24a)

Note that Eqn. (6.24a) reduces to

$$E(N) = \frac{Q^2}{2C_\Sigma} = \frac{1}{2}C_\Sigma V^2$$ (6.24b)

when N is large.

Any electron on the island has an electrostatic potential energy due to the interaction with the remaining N-1 electrons. The variation of E with N is computed by differentiating Eqn. (6.24a) [6.1],

$$U(N) = \frac{\partial E(N)}{\partial N} = \frac{q^2}{C_\Sigma}(N - \frac{1}{2}).$$ (6.25)

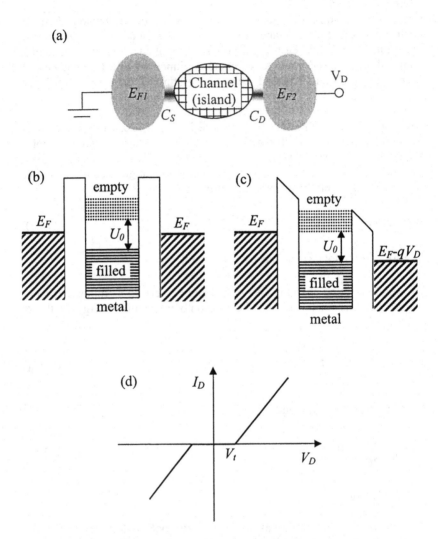

Figure 6.10   Coulomb blockade. (a) The device structure. A small metal island is weakly
coupled to the source and drain contacts. (b) The equilibrium band profile.
Due to the single electron charging effect, an energy gap, U_0, exists between
the highest filled state and the lowest empty state. (c) The band profile at
$V_D=U_0/q$. (d) Sketch of *I-V* characteristics. The source-drain current $I_D=0$ when
$|V_D|<V_t$, which is called Coulomb blockade [6.14].

An electron in a filled state does not feel the electrostatic potential energy due to itself. The potential energy comes from the interaction with the remaining N-1 electrons. For an empty state, the potential energy, however, is due to all N electrons on the island. This difference leads to a single electron charging energy gap between the lowest empty state and the highest filled state even in a metal island,

$$U_0 = U(N+1) - U(N) = q^2/C_\Sigma. \tag{6.26}$$

The equilibrium Fermi level lies in the energy gap. If only a small drain voltage is applied, there are no energy levels between the source and drain Fermi levels, and no current flows. The *I-V* characteristic of the device has a blockade region near V_D=0 as shown in Fig. 6.10d. This phenomenon is called Coulomb blockade. The threshold voltage of Coulomb blockade depends on the position of the Fermi level and the capacitances of the two tunneling junctions. If the equilibrium Fermi level is at the middle of the gap and two tunneling junctions are identical, the blockade threshold voltage $V_t = U_0/q$. As shown in Fig. 6.10c, if a drain voltage $V_t = U_0/q$ is applied, the drain Fermi level moves down by U_0. Because the applied voltage equally drops at the two junctions, the energy levels in the island move down by $U_0/2$. The source and drain Fermi levels align with the lowest empty state and the highest filled state, respectively. The threshold is reached, and increasing the drain voltage slightly leads to current flow.

6.7 Single Electron Transistors

Coulomb oscillations
Figure 6.11b plots the equilibrium band profile at V_G=0 for a single electron transistor as shown in Fig. 6.11a. All states below the Fermi level are filled and the states above the Fermi level are empty. A single electron charging energy gap, $U_0 = q^2/C_\Sigma$, exists between the lowest empty state and the highest filled state, where $C_\Sigma = C_S + C_D + C_G$ is the total capacitance of the island. A zero flat band voltage is assumed, so that the Fermi level is at the middle of the energy gap at V_G=0. As shown in Fig. 6. 11d, the channel conductance at V_G=0 is zero because there are no energy states available at the Fermi level.

The effect of the gate voltage is to modulate the island potential and move the energy level of the island up and down. Based on the capacitance model in Fig. 6.5, a gate voltage of V_G changes the electron potential energy by,

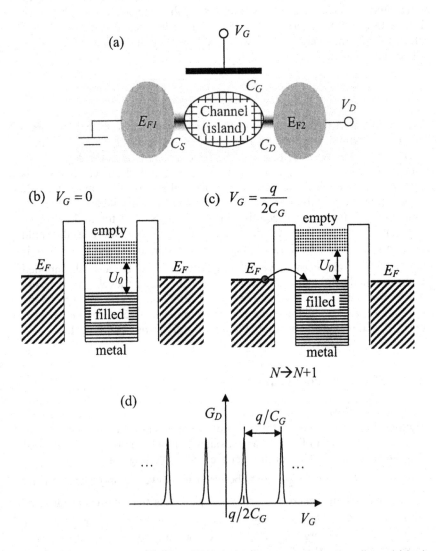

Figure 6.11   Coulomb oscillations. (a) A single electron transistor. A small metal island, which is modulated by a gate, is weakly coupled to the source and drain contacts. (b) The equilibrium band profile at V_G=0. (c) The equilibrium band profile at V_G=q/2C_G. (d) Sketch of the channel conductance vs. the gate voltage.

$$U_{Laplace} = -qV_G \frac{C_G}{C_\Sigma} . \qquad (6.27)$$

A positive gate voltage pushes the energy levels down, and a negative gate voltage moves the energy levels up.

We next plot the equilibrium band profile at $V_G = q/2C_G$. When a gate voltage of $V_G = q/2C_G$ is applied, the energy levels in the island flow down by

$$\left|U_{Laplace}\right| = qV_G \frac{C_G}{C_\Sigma} = U_0/2, \qquad (6.28)$$

and the lowest empty state aligns with the source and drain Fermi level. If an additional positive infinitesimal gate voltage is applied, the lowest empty state is below the Fermi level and both contacts will fill it. One more electron flows into the island, which produces a single electron potential and moves the remaining energy levels up by U_0. The total number of electrons in the island increases from N to $N+1$, and the band profile is shown in Fig. 6.11c. Because there is energy state near the Fermi level, a peak value of the channel conductance is reached, as shown in Fig. 6.11d.

Increasing the gate voltage from $V_G = q/2C_G$ to $V_G = q/C_G$ moves the energy levels in Fig. 6.11c by $U_0/2$, and the band profile becomes the same as that at $V_G = 0$, except that the number of electron in the island is $N+1$ instead of N. The same band profile leads to the same channel conductance at V_G and $V_G + q/C_G$. The periodic oscillations as shown in Fig. 6.11d are called Coulomb oscillations.

Coulomb diamonds
The threshold voltage of Coulomb blockade, V_t, depends on the applied gate voltage. At $V_G = 0$, the equilibrium Fermi level is at the middle of the single electron charging energy gap, as shown in Fig. 6.11b. If a small drain voltage is applied, no current flows because no energy states exist between the source and drain Fermi levels. The I_D-V_D characteristic shows a Coulomb blockade region as shown in Fig. 6. 12. In contrast, at $V_G = q/2C_G$, the equilibrium Fermi level aligns with an energy level as shown in Fig. 6.11c. Applying any non-zero drain voltage results in a no-zero source-drain current. The threshold voltage of Coulomb blockade is zero, as shown in Fig. 6.12.

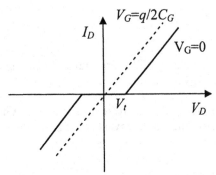

Figure 6.12 I_D-V_D characteristics at V_G=0 and V_G=q/2C_Σ. A Coulomb blockage region with
a threshold voltage V_t exists at V_G=0. At V_G=q/2C_G, no Coulomb blockade
region exist.

The bias points of Coulomb blockade form periodic diamond regions called
Coulomb diamonds on a V_D-V_G plot, as shown in Fig. 6.13. Inside the
diamonds, on current can flow. Next, we derive how the Coulomb blockade
voltage, V_t, varies as a function of the gate voltage, V_G, which determines
the shape of the diamonds.

As shown in Fig. 6.11b, at V_G=V_D=0, the highest filled energy level is at
$E_{HF} = -U_0/2$ and the lowest empty level is at $E_{LE} = U_0/2$. The applied
gate and drain voltage moves the energy levels up and down. For a gate
voltage of $-q/2C_\Sigma < V_G < +q/2C_\Sigma$, the equilibrium Fermi level lies
between the highest filled state and the lowest empty state. The current only
begins to flow when a large enough drain voltage is applied, so that energy
states exist between the source and drain Fermi levels.

Based on the capacitance model as shown in Fig. 6.5, applying a gate
voltage of V_G and a drain voltage of V_D moves the energy levels by an
amount of,

$$U_{Laplace} = -\frac{q}{C_\Sigma}(C_G V_G + C_D V_D).$$ (6.29)

The highest filled energy level moves to

$$E_{HF} = -U_0/2 + U_{Laplace},$$ (6.30)

and the lowest empty energy level moves to

$$E_{LE} = U_0/2 + U_{Laplace}.$$ (6.31)

At a fixed gate voltage, applying a positive drain voltage V_D lowers the energy levels of the island down by $|U_{Laplace}| = qV_D(C_D/C_\Sigma)$, and moves the lowest empty energy level toward the source Fermi level, $E_{FI}=0$. A minimum drain voltage of V_{t1} is needed to move the lowest empty energy level below the source Fermi level,

$$E_{LE}(V_G, V_D = V_{t1}) = E_{F1}.$$ (6.32)

V_{t1} is computed by substituting Eqn. 6.31 into the above equation,

$$V_{t1} = -\frac{C_G}{C_D}\left(V_G - \frac{q}{2C_G}\right).$$ (6.33)

On the other hand, the drain Fermi level lowers to $E_{F2}=-qV_D$ and moves closer to the highest filled state. A minimum drain voltage of V_{t2} is needed to move the drain Fermi level below the highest filled state,

$$E_{HF}(V_G, V_D = V_{t2}) = E_{F2}.$$ (6.34)

V_{t2} is computed by substituting Eqn. 6.30 into the above equation,

$$V_{t2} = \frac{C_G}{C_G + C_S}\left(V_G + \frac{q}{2C_G}\right).$$ (6.35)

Source-drain current begins to flow either when the lowest empty energy level is below the source Fermi level or the highest occupied energy level is above the drain Fermi level. The blockade threshold voltage is determined by the minimum value of V_{t1} and V_{t2},

$$V_t^+(V_G) = \min(V_{t1}, V_{t2})$$
$$= \min\left[-\frac{C_G}{C_D}\left(V_G - \frac{q}{2C_G}\right), \frac{C_G}{C_G + C_S}\left(V_G + \frac{q}{2C_G}\right)\right].$$ (6.36)

Equation 6.36 determines the shape of the Coulomb blockade as shown in Fig. 6.13. When V_G varies from $-q/2C_G$ to $q/2C_G$, the Coulomb blockade threshold voltage first increases with the gate voltage in a slope of $C_G/(C_G+C_S)$, and then decreases with the gate voltage in a slope of $-C_G/C_D$. Notice that for identical source and drain tunneling junctions, the maximum Coulomb blockade voltage is reached as a non-zero gate voltage. When a negative drain voltage is applied, the Coulomb blockade threshold voltage can be computed similarly,

$$V_t^-(V_G) = \max\left[-\frac{C_G}{C_D}\left(V_G + \frac{q}{2C_G}\right), \ \frac{C_G}{C_G+C_S}\left(V_G - \frac{q}{2C_G}\right)\right]. \qquad (6.37)$$

The positive and negative Coulomb blockade threshold voltages define a diamond region in the V_G-V_D plot, within which Coulomb blockade occurs. The diamond region repeats itself periodically, as Coulomb oscillations along the V_G axis.

Experimentally, one measures I_D while varying V_G and V_D. By using color to represent the magnitude of I_D (e.g., red for high current and black for low) a two dimensional plot is produced. Observation of a diamond pattern on such a plot is taken as the experimental signature of Coulomb blockade.

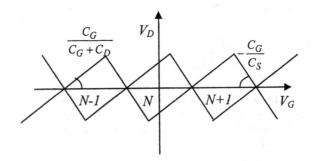

Figure 6.13   Coulomb diamonds. For a SET as shown in Fig. 6.11a, Coulomb blockade occurs for any gate bias V_G and drain bias V_D point inside the diamonds. (After [6.15])

6.8 Summary

This chapter first presented a "general" model for nanoscale transistors, which provides a simple picture for current conduction through nanoscale conductors. The theory not only applies to nanoscale Si MOSFETs with a 2D channel and nanowires/nanotube MOSFETs with a 1D channel as discussed in the previous chapters, but also applies to molecular transistors with a 0D channel as discussed in this chapter, as long as the coupling between the channel and the contacts is strong. Note that in this case, we only deal with an energy-dependent density-of-states, There is no need to think in terms of an $E(k)$ and $+k$ and $-k$ states, as we did earlier. This is important because $D(E)$ can always be defined, but an $E(k)$ is strictly defined only for an infinite, uniform system.

When the coupling is weak and the charging energy of discrete numbers of electrons in the channel (island) becomes important, then a different kind of transistor, a single electron transistor, is obtained. The I-V characteristics of a SET are qualitatively different from a MOSFET. The general model breaks down for a single electron transistor because the discreteness of the electronic charge must be included. (See Ref. [6.1], however, for a discussion of how the general features can be captured by using different self-consistent potentials for spin up and spin down electrons.) The major characteristics of a SET, such as Coulomb blockade, Coulomb oscillation, and Coulomb diamonds, however, can also be easily understood by considering the single electron charging effect.

Chapter 6 References

[6.1]   S. Datta, *"Quantum Transport: Atom to Transisto*r," Cambridge University Press, Cambridge, UK, 2005.

[6.2]   S. Datta, W. Tian, S. Hong, R. Reifenberger, J. Henderson, and C.P. Kubiak, "STM Current-Voltage Characteristics of Self-Assembled Monolayers (SAM's)," *Phys. Rev. Lett.*, **79**, 2530-2533, 1997.

[6.3]   T. Bohler, J. Grebing, A. Mayer-Gindner, H. V. Lohneysen, and E. Scheer, "Mechanically controllable break-junctions for use as electrodes for molecular electronics," *Nanotechnology*, **15**, pp. S465-471, 2004.

[6.4]   W. Wang, T. Lee, I. Kretzschmar, and M.A. Reed, "Inelastic Electron Tunneling Spectroscopy of Alkanedithiol Self-Assembled Monolayers," *Nano Lett.*, **4**, pp. 643-646, 2004.

[6.5]   M. A. Reed, C. Zhou, C. J. Muller, T. P. Burgin, and J. M. Tour, "Conductance of a Molecular Junction," *Science*, **278**, pp. 252-254, 1997.

[6.6]   J. Park, A. N. Pasupathy, J. I. Goldsmith, C. Chang, Y. Yaish, J. R. Petta, M. Rinkoski, J. P. Sethna, H. D. Abruna, P. L. McEuen and D. C. Ralph, "Coulomb blockade and the Kondo effect in single-atom transistors," *Nature*, **417**, pp.722-725, 2002.

[6.7]   N.P. Guisinger, M.E. Greene, R. Basu, A.S. Baluch, and M.C. Hersam, "Room Temperature Negative Differential Resistance through Individual Organic Molecules on Silicon Surfaces, *Nano Lett.*, **4**, pp. 55-59, 2004.

[6.8]   Magnus Paulsson, Ferdows Zahid, and Supriyo Datta, "Resistance of a Molecule," in *Nanoscience, Engineering, and Technology Handbook*, edited by W. Goddard, D. Brenner, S. Lyshevski, and G. Iafrate, CRC Press, 2003.

[6.9]   Supriyo Datta, "Electrical Resistance: an atomistic view," *Nanotechnology*, **15**, pp. S433-451, 2004.

[6.10]   T. Rakshit, G.C. Liang, A.W. Ghosh, and S. Datta, "Silicon-based Molecular Electronics," *Nano Lett.*, **4**, pp. 1803-1807, 2004.

[6.11]   P. Damle, T. Rakshit, M. Paulsson, and S. Datta, "Current-Voltage Characteristics of Molecular Conductors: Two Versus Three Terminal", *IEEE Trans. on Nanotechnology*, **1**, pp. 145-153, 2002.

[6.12]   M. Di Ventra, S.T. Pantelides, and N.D. Lang, *"First-principles calculation of transport properties of a molecular device,"* *Phys. Rev. Lett.* **84**, 979-982, 2000.

[6.13]   P. Solomon and C.R. Kagan, "Understanding Molecular Transistors," in *Future Trends in Microelectronics: The Nano, the Giga, and the Ultra*, edited by S. Luryi, J. Xu, and A. Zaslavsky, Wiley-IEEE Press, 2004.

[6.14]   D. V. Averin and K. K. Likharev, "Coulomb blockade of tunneling, and coherent oscillations in small tunnel junctions," *J. Low Tem. Phys.* **62**, 345-372, 1986.

[6.15]   K. K. Likharev, "Single Electron Devices and Their Applications," Proceedings of the *IEEE*, vol **87**, pp. 606-632, 1997.

[6.16]   K. K. Likharev, "Electronics Below 10 nm". In: *Nano and Giga Challenges in Microelectronics*, ed. by J. Greer *et al.*, Elsevier, Amsterdam, 2003.

[6.17] S.M. Goodnick and J. Bird, "Quantum-Effect and Single-Electron Devices, *IEEE Trans. on Nanotechnology*, **2**, pp. 368-385, 2003.

Index

Page numbers with *italic "f"* indicate figures.